# *The Innovator's Situation*

# *The Innovator's Situation*

*Upper-Middle-Class Conservatism in Agricultural Communities*

*by Frank Cancian*

*Stanford University Press, Stanford, California*

1979

*Stanford University Press*
*Stanford, California*

*Printed in the United States of America*
*ISBN 0-8047-1017-1*
*LC 78-65327*

*To Evon Z. Vogt*
*For good ideas, good advice, and good times,*
*and for creating a space where*
*everyone could be more*

# *Acknowledgments*

The research reported here began in fall 1974. It has been supported by the Center for Research on International Studies of Stanford University, the Ford Foundation, and the Rockefeller Foundation. I am grateful to these institutions and to their officers who worked with me.

The book depends on secondary analysis of data gathered and processed by other researchers for other purposes. Their consistently selfless efforts to facilitate my use of the data provided one of the greatest personal pleasures associated with the project; they made abstract principles of scientific cooperation come alive. Those who helped include: Joseph Alao, Susan Almy, Randolph Barker, Frederick Fliegel, John Gartrell, Peter Gore, Barbara Grandin, Herbert Lionberger, Max Lowdermilk, G. Parthasarathy, Robert Polson, Refugio Rochin, Everett Rogers, and Eugene Wilkening.

At Stanford University, during the beginning of the research, Tom Glenn and Yilmaz Esmer were my research assistants. Both brought great technical skill and personal ingenuity to the work. At the University of California, Irvine, Alice Saltzman was my research assistant during data analysis and writing. Her intelligence, enthusiasm, and ability to handle both theoretical issues and detail improved the analysis and presentation and saved me from errors at many points. I have also been fortunate to have occasional help from a number of others, among whom I want especially to thank Kathy Alberti, Olivia de la Rocha, Robert Sagey, and Lee Sailer.

Many colleagues have been patient and helpful in various ways during the years this research has taken: Francesca Cancian, George Collier, Walter Falcon, Michael Hannan, Charles Lave, Stuart Plattner, and the Brainpickers have been most helpful, and I very much appreciate their efforts.

F.C.

# *Contents*

# *Tables*

**Text**

# *Figures*

**Text**



**Appendixes**

*The Innovator's Situation*

*Chapter One*

# Introduction

This study relates risky, innovative behavior to social position. Its empirical focus is the adoption of agricultural innovations; and information on the behavior of more than 6,000 individuals living in eight countries is used to test the principal hypotheses. Most of those studied are peasants in third-world countries, but many relatively prosperous family farmers in the United States are also included. The conclusions apply to them as well as to the peasants.

The theoretical focus of the study is social stratification. The primary goal bridges the empirical and theoretical foci: it is to assess the relationship of economic rank and innovation. In particular I will question the idea that higher-ranking people are more innovative than lower-ranking people. Both the social nature of economic rank and the identification of the social system in which rank is held prove to be crucial in this effort, and they are discussed at length in both general and specific terms. It will be shown that an understanding of farmers' behavior is greatly enhanced by clarification of a few simple conceptual issues; and that these abstractions can in turn be evaluated in terms of their usefulness in understanding farmers' behavior.

The results are also relevant to assessing the role of information and the role of uncertainty in the adoption of new farming practices. In particular it is found that the relation of rank to adoption changes as the degree of risk and uncertainty change. Since the theoretical focus here is on social processes, they are emphasized. The conceptual and theoretical niceties concerning information, risk, and uncertainty to which the findings are relevant make little or no difference to the discussion of rank and innovation. Thus, they are treated in separate papers (Cancian 1979b, Cancian n.d.).

This chapter highlights some practical implications of the study and some conceptual issues that are useful as orientation to the detail that follows. For a fuller overview of the general issues, the conclusions and implications (Chapter 7) should be read before Chapter 2.

### *Upper-Middle-Class Conservatism and Innovation in Agriculture*

Specifically, I am interested in the relation of a farmer's wealth, or his economic rank within his community, to his inclination to adopt new farming practices. The received wisdom is that larger farmers are more likely to innovate than smaller farmers, that the richer you are the more likely you are to adopt new farming practices.

What follows shows that it is a great oversimplification to say that "wealth and innovativeness appear to go hand-in-hand" (Rogers and Shoemaker 1971: 187). The evidence presented in this book makes it clear that under many conditions the relation of economic rank and innovation within a farming community is more like that shown in Figure 1. That is, the richest or highest-ranking farmers in a community, whether in India, Japan, or Missouri, are most likely to be among the first to adopt, and the poorest farmers are least likely to be among the first to adopt. But, in the broad middle range of farmers who are neither rich nor poor by local standards, the group that might be identified as the "lower middle class" is more likely to adopt early than the group that might be identified as the "upper middle class." I have labeled this pattern "upper-middle-class conservatism." Just when and why it is so, and when and why it is not so, will occupy most of the pages that follow.

Here I want to briefly discuss some practical implications of the principal finding, for part of the justification for all the attention I am giving to the relationship between rank and innovation is that different policies are suggested by the alternative conclusions about its shape. Thus, it seems appropriate to specify some of those differences at the outset.

There are many reasons why current programs designed to

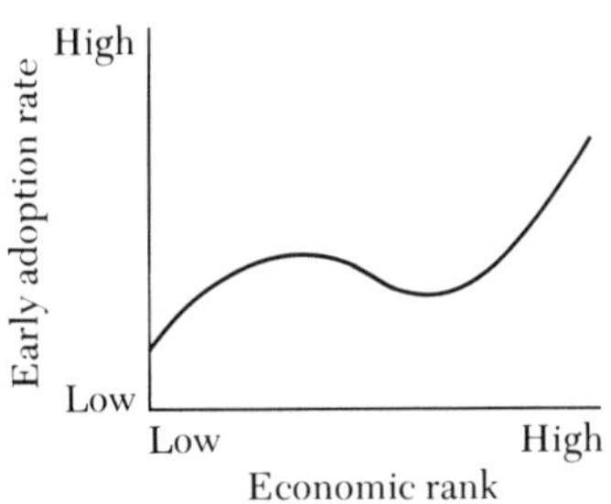


***Fig. 1.*** *The relation of economic rank and early adoption of new agricultural practices*

help the small farmer may end up aimed at upper-middle-class farmers. Principal among them is the fact that, in many less-developed countries, these farmers appear to be the only ones except the rich who have viable commercial farming operations. If such programs fail to induce adoption among the upper middle class, the received wisdom leads to the conclusion that programs and policies designed to give the lower middle class access to new agricultural practices are even more certainly doomed to failure. This follows from the dominant idea that the tendency to innovate increases in an essentially linear fashion with wealth or economic rank.

The results presented below show that the curve displayed in Figure 1, not a positive linear function, is the best description of the relation of wealth or economic rank and the tendency to innovate. Given this fact, the failure of a program designed for what amounts to the upper middle class cannot be taken as evidence that poorer local farmers will resist or reject new practices. Farmers poorer than the upper middle class may still respond very positively to programs appropriate to their scale of operations. In fact, the findings suggest that lower-middle-class farmers are more likely to adopt new practices than are upper-middle-class farmers.

The relation described in Figure 1 and the resultant policy implications stand even after careful consideration of all the complicated details that follow. Nevertheless, it is important to avoid confusion about what is really demonstrated, to clarify the use of terms like “wealth,” “rank,” “early adoption,” and "innovation," and to reconcile the limited evidence brought to bear here with the massive accumulation of studies that support the received wisdom. In addition, to say that the upper middle class is conservative without showing that such behavior is understandable for any person in such a situation is to contribute to stereotyping without increasing the potential for positive social change. This study tries to make it clear that conservatism makes good sense for anyone in the situation of an upper-middle-class person.

### *The Innovator's Situation*

The basic orientation here sees people as similar and their social situations as different. In this study it is the variance in situations that ultimately explains variance in people's behavior. The contrasting approach sees many diverse people in similar situations, and it is the variance in personal characteristics that ultimately explains variance in behavior. People become the carriers of characteristics rather than the occupants of roles and statuses. While it is easy to write and read these simple sentences, it is difficult to keep this distinction and its implications in mind. Thus, I want to briefly explore the meaning of the "situational" approach in general, and its application in the study of the spread of innovation in particular.

In a sense, I am just stating that I will take a sociological approach. Years ago Merton and Rossi said: "That men act in a social frame of reference yielded by the groups of which they are a part is a notion undoubtedly ancient and probably sound. [Sociology] has always been centered on the group determination of behavior" (1957: 234). At the same time they pointed out the lack of productive use of this orientation.

> Thus, half a century ago, DuBois noted that "A white Philadelphian with $1,500 a year can call himself poor and live simply. A Negro with $1,500 a year ranks with the richest of his race and must usually spend more in proportion than his white neighbor in rent, dress and entertainment." But though the specific fact that self-appraisals are *relative* to "the" group framework was often remarked, it was not conceptualized in terms general enough to lead to systematic research on the implications of the fact. (1957: 276, emphasis in the original.)

Merton and Rossi made these observations in the discussion of the reference group concept, which, over the years (Schmitt

1972), has come to specifically refer to groups of which the actor is not a member (that is, nonmembership groups). The more general sociological emphasis taken by DuBois, and underutilized by others, concerns groups or societies of which the actor is a member. This latter orientation is the one followed here.

On the face of it, this approach should be easy to follow. Yet, as the remarks by Merton and Rossi suggest, the basic importance of position within the group or society often slips away. This certainly has been true in the long rural sociology tradition of attention to characteristics of adopters and adopter categories. The tendency to see sociological characteristics as adhering to individuals is especially evident in the first edition of Rogers' famous synthesis of the field, *Diffusion of Innovations*. Age, education, social status, and financial position are included under the heading "personal characteristics" in the description of the characteristics of adopter categories (Rogers 1962: 171–175). By the later edition of the book, these characteristics of adopters are included under the heading "socioeconomic characteristics," which is distinguished from "personality variables" (Rogers and Shoemaker 1971: 185–187). It is also true that the dependent variable in adoption studies is often conceptualized as innovativeness, an individual characteristic. On the other hand, Lionberger, in his earlier general book on the adoption process, lumps age, education, and "psychological characteristics" under "personal factors" and calls farm income, size of farm, and tenure status "situational factors" (1960: vi).*

The tradition of diffusion research is presently being criticized because of its focus on the individual (Goss 1977), but a principal active alternative is critical focus on corporate farming and what is called "the structure of agriculture." In the long run I fear this new emphasis may end up as a focus on big "individuals" as compared with small individuals. While this focus is crucial to currently important social criticism, it is in danger of shifting the old template to corporate agriculture. This would produce knowledge about the characteristics of corporations that is parallel to the knowledge about the characteristics of individuals produced by the diffusion research tradition that is being criticized. I hope the structure of agriculture approach

* Lionberger's use of "situation" in a way that is, in general terms, similar to the use intended here, goes back to his earlier publications on Missouri farmers (1948, 1952).

will concentrate on the position of corporations within the larger economic system.

In this book I attempt to avoid these difficulties. I will concentrate on (relative) rank and give attention to the sociologically meaningful limits of the society or group within which rank is held. These two elements, rank and "community of reference," are at the core of this look at the innovator's situation.

A number of the substantive and methodological implications of the situational approach used here are illustrated by the single application pictured in Figure 1. For example, it will become apparent that the people at the top and bottom of the community ranking scale are influenced by their proximity to the ends of the scale. Thus, we need to know the community of reference so that we can identify the ends of the scale. This need contrasts sharply with the requirements of a "linear" theory, for a straight line should be sliceable at any point without significant diminution of the predicted effects. Thus, identification of the ends is less important. In addition, because of the curvilinear relations between the major variables, certain ordinal measures and crosstabular statistical analysis prove themselves preferable to interval measurement and regression analysis, which are conventionally considered more powerful. The scale of the variables becomes an important substantive and methodological issue.

On the policy side the situational approach has immediate general implications. First, it "blames" the situation, not the victim. Thus, it is not supportive of policies that seek to improve the character or characteristics of some part of the population in the (implicit) hope that everyone, or almost everyone, can be brought above some poverty line. This is so because the situational approach emphasizes the inherently relational nature of many important phenomena. This stress on the relational also suggests pessimism about widespread improvement of individual situations without change of the system.*

### *Some Previous Uses of "Situation"*

"Situation" has been used in a number of senses in social science over the years. Inevitably, each subsequent use bor-

*For a discussion of some of these issues in relation to industrial societies, see Cancian 1979a and Hirsch 1976.

rows a bit from the previous ones. While I cannot survey past uses or fully explore the complex of interrelated meanings, I do want to review some basic meanings that help to clarify my purpose in this book. The reader who is not interested in the intricacies of social science terminology should skip to the last paragraph of this section.

Recently in psychology and anthropology the term has been associated with withdrawal from trends towards rigid monodeterministic approaches. That is, it has been associated with "the recognition that complex human behavior tends to be influenced by many determinants and reflects the almost inseparable and continuous interaction of a host of variables . . ." (Mischel 1977: 246). Mischel ends this sentence with the words "in both the person and the situation" because he is a psychologist dealing with the implications of his own findings that personality traits as commonly measured cannot be taken as good predictors of behavior across situations.

A parallel interaction between the complexity of human behavior on the one hand and simple isolatable principles on the other is found for anthropology in van Velsen's characterization of the relation of structural analysis and situational analysis:

> The "structural frame of reference," according to Fortes (1953, p. 39), "gives us the procedures for investigation and analysis by which a social system can be apprehended as a unity made up of parts and processes that are linked to one another by a limited number of principles of wide validity in homogeneous and relatively stable societies." (van Velsen 1967: 131.)

The extended case method and situational analysis promoted by the Manchester school (for example, Mitchell 1956, Turner 1957, van Velsen 1964, with Gluckman as leader) attempts

> to show how the unique, the haphazard and the arbitrary are subordinated to the customary within a single, if changing, spatio-temporal system of social relations . . . to show how the general and the particular, the cyclical and the exceptional, the regular and the irregular, the normal and the deviant, are interrelated in a single social process. (Turner 1957: 328, quoted by van Velsen 1967: 148.)

In both the psychological and anthropological approaches there is a "contextualism" in which the established variables are seen as too simple to survive the complexity of the "actual" situations. Any general abstract rule or basic relationship of variables is seen to operate in the context of less abstract and less general rules and local peculiarities; and these contextual features, about which no generalizations are asserted, are seen as crucial to the behavioral outcomes. "Situation" used in this contextual sense represents a withdrawal from abstraction and explicit generalization.

In sociology and social psychology the symbolic interactionists have a long tradition of attention to "definition of the situation." Their usage has aspects of both: 1) the contextualism seen in the psychological and anthropological uses discussed above, and 2) a focus on the actor's point of view in both relational and cultural terms. Here I would like to use part of the latter aspect and none of the former.

In sum, I hope to use the word "situation" to emphasize the need to avoid individualistic explanation in terms of personality traits, and similarly to emphasize the need to attend to the local definition of the relevant other people. I do not want the word for its suggestion that context and local detail are important. While this is true, in this study I want to generalize. Thus, the use of "situation" to suggest that things are complicated will not help.

### *Issues of Procedure and Style*

The chapters that follow are organized in a traditional way. Chapter 2 states the theory. Chapter 3 describes the data available to test the theory. Chapter 4 discusses measurement of the independent variable, and Chapter 5 does the same for the dependent variable. Chapter 6 presents the tests of the hypotheses; and Chapter 7 states the conclusions and implications of the study. A number of appendices amplify and document complexities encountered along the way. This organization is satisfactory because it gives a clear outline to the analysis and fits with the relatively brief discussion of general issues in the first and last chapters. I find it preferable to discuss general issues along the way, as the concrete problems of analysis are available to illustrate them.

Before beginning, I want to express the anthropologist's lament: I am torn between the details of each case and my effort to generalize across the world that is represented here by data from many countries. There is no solving this problem. From my own fieldwork in Mexico I happen to know the kinship position and personal situation of a half dozen or so of the hundreds of individuals who end up classified as high middle rank in this study. These characteristics of their local social and personal situations suggest alternative explanations of the behavior patterns predicted by the general theory developed here. As I see it, these alternative explanations add to, but do not detract from, the understanding offered by my general theory.

## Chapter Two

# Rank and Innovation: The Theory

Why is the upper middle class conservative? The preview of conclusions given in Chapter 1 shows an unusual dip in the curve describing the relation of rank and innovation. Both low-middle-rank farmers and high-rank farmers seem to innovate more than the high-middle-rank group. In this chapter I will lay out the theory which explains this unusual pattern and specify the hypotheses that will be tested in Chapter 6.

The theory presented below has a history. It has been stated in relatively formal terms in two places (Cancian 1967, 1972); and that formal version has been subjected to thorough comment and criticism (Gartrell, Wilkening, and Presser 1973; Morrison 1973; Morrison, Kumar, Rogers, and Fliegel 1976; and Gartrell 1977). The formality of the original statements produced implications that might otherwise have been missed, and the comments and criticisms led to significant modifications that are noted in this and subsequent chapters. On the whole it seems pointless to burden the reader with yet another formal exposition. What follows in this chapter is thus in many ways less formal and less detailed than earlier statements of the theory. This statement, like previous ones, is meant as a general theory relating rank to risk of resources relevant to gaining and maintaining that rank; but, as before, it is often easier to express the basic ideas in concrete terms directly relevant to the agricultural innovations that are the empirical focus of this study.

### Basic Considerations

At the outset I will assume that it is in the nature of stratification systems for any individual to prefer high rank to low rank. Most of the theory flows from this proposition and from the idea that the possibility of achieving higher rank is often the motivation for innovation. The particular shape of the relationship between rank and innovation derives from specification and modification of these basic ideas and the conceptions of rank and innovation themselves.

Rank relative to other people in the stratification system indicates control of resources; economic rank indicates control of economic resources. Innovation, as it is defined here, involves the investment of resources in a situation where the return to investment is not certain. In common-sense terms, the innovator, almost by definition, takes a risk, for he or she does not know exactly how the practice will work. This theory concentrates on innovations involving risk of the very resources on which the ranking system is based. For the study of agricultural innovation this translates very simply into the risk of money on new practices that might produce money enough to raise the actor's economic rank.

Both rank and innovation involve important conceptual and definitional problems. For rank these include both the definition of the social system within which rank is seen to operate, and the relative nature of rank within that system. These issues are covered in Chapters 3 and 4 respectively. For innovation the major problem concerns the distinction between the emphasis on information diffusion in the rural sociology tradition and the emphasis on uncertainty in the approach developed here.* Chapter 5 is devoted to innovation. In all cases, the delay of the conceptual discussion until we are more familiar with the concrete cases simplifies that discussion. And in no case does the delay hamper the exposition of ideas in this chapter.

The basic theoretical elements are set out in the next three sections: The Inhibiting Effect of Rank, The Facilitating Effect of Wealth, and The Curvilinear Effect. These abstract elements are then combined to make predictions from the overall theory; and the predictions are briefly illustrated and discussed. Since the exposition proceeds from the very abstract to the very con-

*In this discussion the concepts "risk" and "uncertainity" are not distinguished. The implications of this study for potential distinctions between them are discussed in Cancian 1979b.

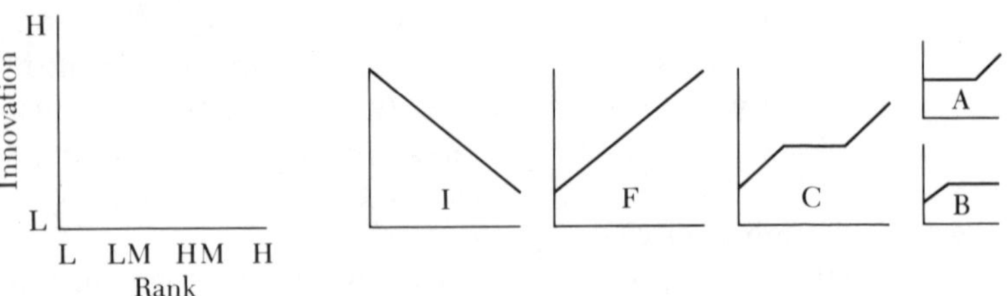


*Fig. 2. Inhibiting, facilitating, and curvilinear effects in the abstract*

crete, those who prefer the opposite order may want to look at the final section of the chapter before continuing.

### The Inhibiting Effect of Rank

The first element in the theory is the idea that high rank inhibits innovation because people of high rank have more to lose and less to gain from a random change in rank. That is, for them there are more ways to lose than to gain rank. At the outset, when little is known about the results a new practice will produce, it makes sense for the high-ranking person to avoid the risk. When uncertainty is high, high-ranking people should seek to maintain their rank while low-ranking people should seek to gain rank. For low-ranking people, a random change in rank is more apt to be for the better.

This specification of the basic considerations leads to the conclusion that, all other things equal, persons of higher rank will innovate less than persons of lower rank. This is the first element in the overall theory of the relation of rank and innovation. It is illustrated as the I curve in Figure 2. The inhibiting effect of high rank is only one of the things contributed by rank to the behavior of innovative and noninnovative actors. Actual behavior cannot be predicted until we have added two more elements to the theory.

### The Facilitating Effect of Wealth

On the other hand, wealth facilitates innovation. Innovations cost money,* and the wealthier you are the more likely it is that

*In general terms they involve the investment of resources relevant to rank.

you will have the money needed for any given innovation. Few innovations are perfectly divisible, and, in the long run, the wealthy farmer will innovate more simply because he can afford to do so. Even if he is less inclined to risk a proportion of his resources (say 10 percent) on innovations, he may end up trying more innovations simply because many more possibilities will come across the budget threshold determined by his wealth; this argument is represented by the F curve in Figure 2.

This facilitating effect of wealth is augmented by the association of wealth with information and education. Wealthier people tend to be better informed. And, since the uncertainty that stems from lack of information inhibits innovation, it makes sense to conclude that better informed people will innovate more in cases where the innovation will be to their advantage. Since this study concentrates on "successful" innovations we may expect the information effect to facilitate innovation.

The F curve also represents the traditional empirical findings described in Chapter 1 and the ideas usually used to explain them. In this formulation it is easy to see that the I and F curves highlight the basic conflict between the inhibiting-effect interpretation and the received wisdom on the relation of rank and innovation. Part of the conflict will be resolved in later sections of this chapter; but part will, of course, remain to be decided by the data presented in Chapter 6.

Notice that this section has been labeled The Facilitating Effect of *Wealth*, while above I referred to the inhibiting effect of *economic rank*. This contrast between economic rank and wealth is meant to call attention to the different underlying principles involved.* The inhibiting-effect argument emphasizes the relative position of people in a rank structure and their response under uncertainty. The facilitating-effect argument emphasizes the actual limits on resources invested even under conditions of certainty. These basic contrasts between rank and wealth, and between uncertainty and secure knowledge, remain central throughout this book.

*For many purposes wealth and rank are interchangeable. High wealth rank is equivalent to relatively great wealth, and vice versa. Chapter 4 reviews situations in which it is productive to maintain the distinction between wealth and wealth rank.

### *The Curvilinear Effect*

The overall theory predicts that the empirical relation of rank and innovation will be curvilinear (see Figure 1). Yet, both the inhibiting effect and the facilitating effect are linear. Here I want to present the various arguments for curvilinearity: 1) the essentially sociological notions that were presented in the original statements of the theory (Cancian 1967, 1972); 2) an argument based on personal characteristics that I will review and reject as an appropriate explanation of curvilinearity; and 3) the argument about characteristics of the normal distribution (presented by Gartrell, Wilkening, and Presser 1973). They will be reviewed in turn.

If we see a person's inclination to risk as a balancing of what he has to gain and what he has to lose, it is not difficult to imagine that people at the ends of the rank continuum might not operate according to the principles that predict the behavior of those in the middle of the continuum. Those at the end have everything or nothing to gain, everything or nothing to lose. When this is the case, one might argue, an economizing conceptualization of the situation does not make sense; those who define the ends of ranking continua may not participate in the competition for rank in the same way as those who are not so conspicuous.

Of course, the question of how they will behave still remains open. I argue that they will behave *unlike* the prediction made for them by the inhibiting effect ideas discussed above. The behavior most unlike the predictions of the initial proposition would be for the highest-ranking people to be high riskers and the lowest-ranking low riskers. These principles yield a relation of rank and innovation like the curvilinear one shown in Figure 2, curve C. Note that in order to illustrate this idea a minimum of four ranks must be distinguished. For convenience, I will refer to these ranks as low, low middle, high middle, and high.

The kind of thinking that goes into arguments for the distinctiveness of people at the ends of the rank continuum is illustrated by George Homans' chapter "Status, Conformity, and Innovation" (1961). Homans usually divides the social continuum into three parts: high, middle, and low. He argues that highs are innovative because they need to maintain their distinctiveness. Lows are innovative because they are not highly rewarded by the group for any of their behavior; thus they find the cost of nonconformity to be more consistent with their expectations than do the middles, who are accustomed to rewards for their conventional behavior. This produces a U-shaped curve.

Homans finds a different interpretation necessary to handle the results of an experiment on conformity to group norms by Dittes and Kelley. In their study lows and very lows were distinguished, making four ranks in all; and the very lows were extremely conformist in their publicly expressed opinions. Here, Homans accepts the interpretation made by Dittes and Kelley: "In the extreme case, however, where acceptance is so low that actual rejection is presumably an imminent possibility, anxiety about rejection is especially high, and the result seems to be a pattern of guarded public behavior" (Dittes and Kelley 1956: 106).

There are essentially two arguments for why high-ranking people are innovators: 1) they are secure in their positions and take risks out of what amounts to boredom (as contrasted with economic motivation); and 2) they realize that their distinctiveness is based on leadership in economic techniques, and they take self-conscious risks in order to maintain this distinctiveness. The usual arguments for why low ranking people are not innovators are also two: 1) they are so poor that any risk threatens "total economic extinction," therefore they are unusually conservative (the Dittes and Kelley point); and 2) they refuse to compete in the economic system because past failures have made it seem like an inefficient way to seek rewards. These arguments are all sociological. That is, they depend on the actor's relative position in the social system. And, they are curvilinear. That is, they suggest why one end of the continuum should be different, while allowing for other (presumably linear) effects for the remainder of the continuum. Together they bend both ends of the continuum and form the A and B effects illustrated in Figure 2.

An argument based on personal characteristics can also be used to explain these phenomena. On the one hand, it is argued that people who have innovation-producing personalities

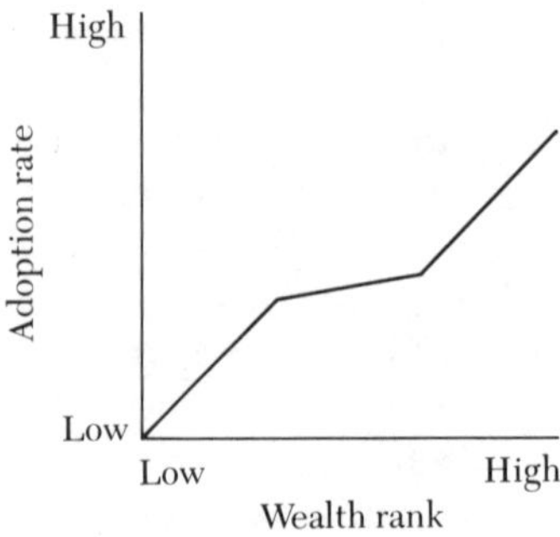


***Fig. 3.*** *The relation of wealth rank and adoption rate when wealth is normally distributed and linearly related to adoption rate*

achieve high rank because of this character trait and remain innovators once they have achieved high rank. On the other hand, it is argued that people who have conservative personalities lose out in the competition for rank because of this character trait and remain the same once they are low. The appropriateness of this argument as an explanation of curvilinearity depends on the curvilinear distribution of personality traits in the population (and their constancy within people).

This argument is tautological unless there is independent evidence for such a distribution. The overall approach taken in this study does not seek support for personality explanations; consequently, no material relevant to the testing of this explanation is included below.

Gartrell, Wilkening, and Presser (1973: 400–402) point out that normal distributions of population characteristics combined with underlying linear scales for both variables might be expected to produce something like the curve in Figure 3 under certain conditions even if none of the substantive relations discussed above hold. That is, they suggest that the curvilinear effect might be an artifact of distributions, not a substantive relationship. For example, assume a linear positive relation of innovation and wealth and a normal (bell-shaped) distribution of wealth; break the normal distribution into quartiles (by number of actors); and plot the quartiles as in Figure 2. The results will have the characteristics of the curve in Figure 3. That is, the middle of the normal distribution will have many individuals who are very similar in the amount of wealth they hold, and thus the dependent adoption rate variable should be expected to increase relatively slowly through the middle of the continuum.

This explanation of curvilinearity has the virtue of simplicity. It does not, of course, produce the negative relation shown in the overall prediction illustrated in Figure 1. Since the dip in the empirical relationship of rank and innovation is the theoretical and practical heart of the theory, the principal argument of the theory is not affected by the Gartrell, Wilkening, and Presser argument.

Nevertheless, the Gartrell, Wilkening, and Presser paper makes problematic the specific arguments for curvilinearity presented above. If their characterization of the quasiartifactual nature of the curvilinearity is correct, the theory explaining the overall relation of rank and innovation might be reduced to the inhibiting effect and the facilitating effect plus the normal distribution of the independent variable.

There are two problems with such an extreme application of Occam's razor. First, as we will see in Chapter 4, conventional metrics produce a positively skewed, not a normally distributed wealth variable in most real populations. Second, the Gartrell, Wilkening, and Presser argument depends on the logic of interval scales, and the basic logic of the theory presented here is that of ordinal scales (for details see Chapter 4). That is, they argue amount of wealth, while the inhibiting effect and the curvilinear effect are otherwise explained in terms of rank order conceptions. Since there is no disagreement about the essential empirical validity of the curvilinear effect, and since the Gartrell, Wilkening, and Presser arguments do not affect the essential predictions of the overall theory, these matters will not be discussed further here. I do want to note that 1) a number of hypotheses presented in my original statements (1967, 1972) of the theory have been dropped in response to Gartrell, Wilkening, and Presser's argument, and 2) the relation of their argument to future testing of those hypotheses is not fully explored here.

### *Predictions from the Overall Theory*

The three abstract elements presented in the complex of arguments considered above can be brought together to yield predictions about the empirical relation of rank and innovation. Insofar as the inhibiting, facilitating, and curvilinear effects present an adequate overall theory of the relation of rank and innovation, any curve representing an empirical case of this relation must be recognizable as a combination of the I, F, and C curves. Since curves I and F negate each other, the inhibiting effect and the facilitating effect will never be evident in the same descriptive curve, although they may well be present in the same empirical situation. Thus, any empirical curve must be either a C curve (when I and F exactly negate each other) or a combination of I or F with C. The possible combinations are illustrated in Figure 4 and discussed below.

The predictions will concentrate on the inhibiting and facilitating effects. Those effects are of central theoretical and practical importance in this study. As noted above, the facilitating effect represents the received wisdom described in Chapter 1; and the inhibiting effect represents the attempt to modify that approach through emphasis on relative position and uncertainty. Predictions that illuminate how and when these effects operate will be most useful both for understanding the relation of rank and innovation and for administering the spread of innovations. The curvilinear effect remains as a part of the empirical reality that will be found in most studies of actual behavior. As we saw above, the arguments interpreting the behavior pattern described by the curvilinear effect are complex and ambiguous. Since, in the present context, they are also less important, they will not be used to develop further predictions here.

This means that the middle of the rank continuum will be the focus of our attention. This is because the curvilinear-effect argument makes no claims about the relationship of rank and innovation in that part of the rank continuum. Thus, the inhibiting effect and the facilitating effect sharply confront one another in that part of the continuum. In terms of the four-part rank continuum described above in the discussion of the cur-

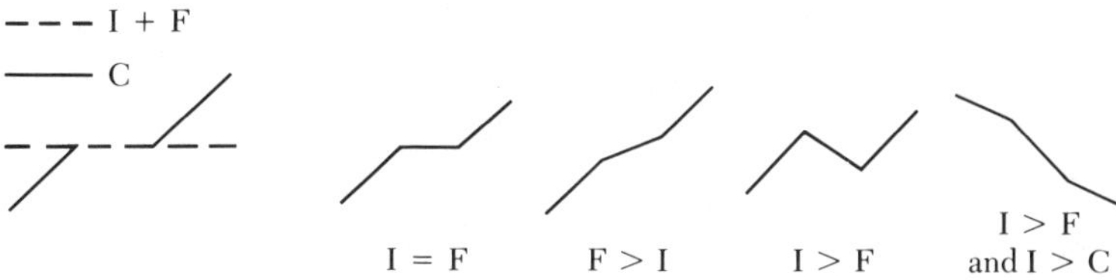


***Fig. 4.*** *Possible empirical combinations of the inhibiting, facilitating, and curvilinear effects*

vilinear effect, what follows will focus on the innovation rates of the lower middle and the upper middle ranks. In developing and illustrating these predictions, I will usually assume that the curvilinear effect will, in some degree, hold. That is, I will not consider the option illustrated in Figure 4 by the I > F and I > C curve.

Note that the overall theory makes no specific prediction about the dominance of the inhibiting effect or the facilitating effect. The only clear prediction is that the strength of the inhibiting effect should decrease as uncertainty decreases. That is, early in the innovation process, when uncertainty about the outcome of using a new practice is high, the inhibiting effect should be at its strongest. As more and more people adopt an innovation, information spreads, uncertainty is reduced, and the inhibiting effect should gradually disappear. As the new practice becomes an old practice, as the innovation becomes a secure routine, the inhibiting effect should diminish. Since the inhibiting effect and the facilitating effect "contradict" each other, the disappearance of the inhibiting effect means the emergence of the facilitating effect in the empirical curve.

If the inhibiting effect never dominates the facilitating effect it would be impossible to distinguish the combination of the F and C curves from the combination of the F, C, and I curves. That is, F > I in Figure 4 looks no different from what F + C alone might look like. Thus, if the inhibiting effect can be observed in actual behavior, it must dominate at some point at least in the way illustrated in the I > F curve in Figure 4. According to the theory developed here, this situation is most likely to occur in the early stages of the spread of an innovation.

Thus, the first hypothesis is:

HYPOTHESIS A: *In the early stages of the spread of an innovation, low-middle-rank individuals are more likely to adopt it than are high-middle-rank individuals.*

Since the spread of an innovation and the concomitant increase of information will tend to decrease the strength of the inhibiting effect relative to the facilitating effect, we may also predict that high-middle-rank people will increase their adoption rate relative to low-middle-rank people as the innovation spreads:

HYPOTHESIS B: *In the later stages of the adoption process, the adoption rate of high-middle-rank individuals will increase relative to the adoption rate of low-middle- rank individuals.*

In other words, in later stages of the adoption process the overall or cumulative adoption rate of the high middle rank should tend to catch up with and even pass the overall or cumulative adoption rate of the low middle rank.

These two hypotheses provide tests for the fundamental ideas in the inhibiting-effect-of-rank explanation of upper-middle-class conservatism. Hypothesis A tests whether the effect of rank inhibition is important enough to show empirical dominance at some stage of the innovation process. Hypothesis B tests whether the crucial role attributed to uncertainty leads to any useful distinction between earlier and later stages of the adoption process. While these ideas are at the core of the present study, the study is more than a simple test of them. Organization of the inquiry so as to test them will require many other reevaluations of old ideas and old data.

### *Illustration and Discussion: The New Theory and the Old Findings*

A concrete illustration may help the reader to retain the complex ideas reviewed above. The one I offer here also serves to clarify a number of issues, including the apparent contradiction between the predictions of the theory developed here and the well-established findings described in Chapter 1. It seems unlikely that a new theory, however engaging its fundamental ideas, could undo the weight of evidence from so many studies.

TABLE 1. *Predicted Distribution of Adoption by Rank: An Illustration*

| Adoption | Low rank | Low middle rank | High middle rank | High rank | Total |
|---|---|---|---|---|---|
| Remainder | 80 | 55 | 45 | 20 | 200 |
| Stage 2 | 10 | 15 | 35 | 40 | 100 |
| Stage 1 | 10 | 30 | 20 | 40 | 100 |
| TOTAL | 100 | 100 | 100 | 100 | 400 |

The compatibility of the new theory and the old findings, as it is illustrated below, enhances the value of both of them because it more carefully specifies when each is most likely to be relevant.

To illustrate, let us suppose that we have studied a community in which all 400 household heads are full-time farmers who own the land they farm. An innovation (say, hybrid corn seed) has been adopted by all of them by the time of the study, and the year of first use is known for each farmer. Both the adoption/innovation variable (year of first use) and the rank variable (size of operation) are continuous and permit arbitrary division into the stages and ranks shown in the totals column and row of Table 1. That is, 100 farmers are placed in each of the four ranks necessary for the analysis; and the first 100 farmers to adopt the innovation are taken as Stage 1 adopters. In Stage 1 uncertainty is higher than it is in Stage 2, which is operationalized as the second 100 farmers to adopt the new corn seed.

The entries in the cells may be read as actual numbers of farmers or as column percentages. Note that the figures for Stage 1 confirm Hypothesis A. That is, the low middle rank has a higher adoption rate than the high middle rank. The dip described in Chapter 1 is closely approximated by this imaginary data. Note also that Hypothesis B is confirmed. That is, in Stage 2 the adoption rate of the high middle rank goes up relative to the adoption rate of the low middle rank.

An important conclusion about the relation of the new theory and the old findings is most clearly illustrated by the graphic version of Table 1 that is presented in Figure 5. Note that cu-

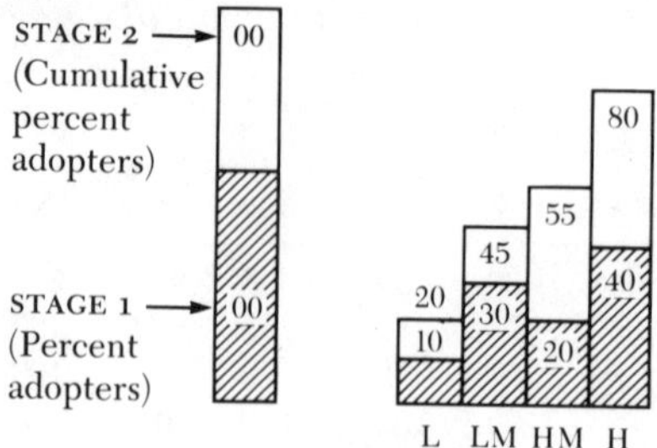


***Fig. 5.*** *Adoption rate by stage: The new theory and the old findings*

mulative adoption by the end of Stage 2 is monotonic. In fact, simple adjustments in the figures would make it linear without any damage to confirmation of either hypothesis. That is, under the theory developed in this chapter, the dip that is so crucial an element in the relation of rank and adoption in the earlier stages of the spread of an innovation may disappear and be completely obscured by the later stages of the adoption process (the size of the stages is not specified by the theory). If this is so, then it may be possible to explain the traditional findings reviewed in Chapter 1 as the results of analyses that did not distinguish stages. At the same time, the dip showing upper-middle-class conservatism may remain in the early stages of the spread of an innovation.

Thus the old findings and the new theory are perfectly consistent with one another. At the same time, the important contradictions between the old theory and the new theory, and their respective policy implications, remain.

Deciding which theory best explains observed adoption behavior, and under what conditions, remains as the task of subsequent chapters. There we will also explore the myriad details involved in deciding which of the many possible operationalizations of the overall theory is most appropriate, given the realities of farming and the realities of available data on farming.

*Chapter Three*

# Communities and Data Sets

The theory stated above is meant to apply to all farmers and all farming communities. Thus, a proper test of it must encompass data on a variety of farming communities in a variety of settings. Information is also needed on the economic rank and innovative behavior of a large number of farmers in each community. Such data on rank and innovation in a variety of communities can be used both to test the theory and to indicate where it does not apply.

The identification and collection of appropriate data is a major undertaking for two reasons. First, it requires interpretation and specification of the theory to make it fit real world situations and the practical realities of data collection. Second, the need for data on many farmers in each of many different communities simply generates a volume of detail that is difficult to manage.

This chapter lays the groundwork for the empirical part of the study. It begins with descriptions of the available data and the collection process and goes on to details about the collected data sets and the populations they represent. Finally, it provides a preliminary discussion of the community of reference notion and its relation to the assembled data sets.

## *The Available Data*

A truly comparative study involving farmers and farming communities in a wide variety of settings would be impractical for a single researcher or research team. Too many places, too many technologies, too many languages, too many complications would be involved. In addition, the relatively limited question being asked in the present study would not justify the expense involved in collecting data in a variety of communities. Fortunately, over the past three decades, rural sociologists and other social scientists have studied hundreds of communities all over the world. In many cases they have collected data very close to those required to test the theory developed in Chapter 2.

Most of these studies are included in the more than 1,000 citations in Rogers and Shoemaker, *Communication of Innovations* (1971). The authors summarize the study of diffusion of agricultural innovations in the rural sociology tradition. This research tradition began in United States land grant colleges before World War II with such classics as the Ryan and Gross study (1943) of diffusion of hybrid corn seed in two Iowa communities; and it grew substantially in the postwar years. In 1960 Lionberger cited 100 research studies in his *Adoption of New Ideas and Practices*; and Rogers' original edition of *Diffusion of Innovations* (1962) was based on 405 studies. In the decade of the 1960s especially, diffusion of innovation research was internationalized. Hundreds of parallel studies in less-developed countries were done by United States-based and native researchers (see Rogers 1976). Thus, the potentially available studies of the spread of agricultural innovations are, from a practical point of view, infinite in number.

On the other hand, not all these studies cover the variables of primary concern here. Not all of them meet the other conditions most important for testing the theory presented above. Not all of them have the sample size and selection appropriate to the present effort. And not all of them are easily available from the original investigator.

### Identifying, Collecting, and Selecting the Data Sets

The process of going from 1,000 citations to the roughly 20 data sets used here will be described as involving three stages: identification, collection, and selection. Simple doggedness dominated the process. It was relieved only by the selfless cooperation of colleagues who provided the data and by the excitement of receiving each new usable data set.

The identification of potentially relevant studies involved two main efforts. First, with the help of a research assistant, I undertook to explore the 227 studies that Rogers and Shoemaker list as relevant to this generalization: "Earlier adopters have larger sized units (farms, and so on) than do later adopters" (1971: 361). We began here because all these studies had at least collected data on both economic rank and adoption of innovations. Second, I wrote to a number of rural sociologists and others who had been helpful when I did my earlier study on a similar question (1967), and asked them if they had access to or knew about any relevant data sets. My letters to these people, and to authors of studies identified in the literature, included the data characteristics sheet reproduced here as Figure 6. I tried to interpret the ideals stated on the sheet loosely so as to maximize the number of appropriate studies. A few times, limited resources made me cut off the investigation before I was absolutely certain that the study in question was not usable. Without doubt some appropriate studies were missed.

This procedure insured a large initial universe of studies that in some way measured the economic rank and the innovation variables. Within this universe there was no attempt at sampling. I tried to collect all the appropriate data sets, and no data set that met the criteria of appropriateness was ever rejected. This is as much as can be said about the sampling of the communities studied here.

Collecting or assembling the data involved a variety of contacts, most by mail and telephone, with original researchers who were asked to share data. Tom Glenn made one trip on my behalf to the University of Michigan and Michigan State University, where he was aided by Everett Rogers in the use of the Diffusion Documents Center and the data banks; and Herbert Lionberger helped me to clarify some points during a trip I made to the University of Missouri for other purposes. Data came in various forms, from tabulation sheets and computer listings to computer cards and tapes. Along with the data, codebooks and publications were collected. The colleagues who generously provided the data were also uniformly cooperative in helping to clarify ambiguities in codebooks that were in some cases very old.

## *Characteristics of Data for Economic Rank and Adoption Study*

### *The Sample*

Two hundred or more farmers living in the same community.

Populations whose members live in different physical communities may be used insofar as they form a reference group.

For populations that include many part-time farmers, it will be useful to separate out full-time and part-time farmers and to concentrate on analysis of full-time farmers.

For populations that include many different types of farming (e.g., grains and dairy) it will be useful to separate the major types of activity.

### *Economic Rank Variable*

Acreage, number of animals and gross value of annual produce will be taken as very good measures of this variable.

Level of living, tax reports, ratings by alters, etc., may also be used.

While data providing a continuous distribution on this variable would be ideal, any data permitting division of the population into approximate quartiles will be usable.

### *Adoption Variable*

Time (year) of adoption of a crucial innovation will be the ideal measure of this variable.

Indices based on time of adoption of many innovations or on the number of innovations adopted by a given point in time may also be used.

For indices that include expensive and inexpensive innovations, it will be useful to be able to separate innovations on the basis of cost.

While data providing continuous distribution on this variable would be ideal, any data permitting the division of the population into approximate quartiles will be usable.

### *Other Variables*

It will be useful to be able to control for age, education, types and number of information sources and other variables that are often important to adoption.

Frank Cancian
Department of Anthropology
Stanford University
Stanford, California 94305
October 1974

*Fig. 6. Sheet used in search for data*

Not every data set collected was used. Of the half dozen or so sets not used, most were rejected because they did not include adequate numbers of farmers within a single community of reference, and a few were rejected for inadequate variance on the dependent variable. On the whole, the data sets rejected after collection had been collected in an effort to complete the record in a set of parallel studies, or had "come along with" some more appropriate data set. Those collected and then rejected include the Brazil case from the three-nation study (Rogers et al. 1970) and the second phase of the study that appears here as India 412.

Appendix C discusses cases that were included even though they did not meet the criteria stated in Figure 6.

### *The Data Sets*

The twenty-three data sets listed in Table 2 provide the empirical basis for this study. Table 2 and a number of parallel tables in this and subsequent chapters include the greater part of the basic facts about each data set. This chapter focuses on the characteristics of the community and the sample, Chapter 4 on the economic rank variable, and Chapter 5 on the innovation variable and the types of farming involved.

Table 2 identifies the data sets. It lists the country and state or parallel political unit in which the study was done, the date of field research, the number of individuals in the sample analyzed here (the N column), and one publication as a major source on the study. The N column is carried from table to table to permit identification of data sets carrying the same country name. Unfortunately, there is no single less awkward and ugly characteristic of the data set that will do this job. The publication provides a positive identification in the literature and opens access to details provided in the reports of the original researchers. A more extensive list of background publications is found in Appendix D. Notes on the references and data sources columns of Table 2 indicate who provided the actual data set used in this study. A note on the country name indicates subsequent inclusion in a public data bank from which it is available to other researchers.

TABLE 2. *Identification of the Original Studies*

| Country | N | State or province | Date of fieldwork | Reference and data source [a] |
|---|---|---|---|---|
| 1) India | 185 | Andhra Pradesh | 1972 | Parthasarathy 1975 |
| 2) India | 412 | Uttar Pradesh | 1963–64 | Roy et al. 1969 |
| 3) Kenya | 540 | Eastern | 1972–73 | Almy* 1974 |
| 4) Mexico | 108 | Chiapas | 1966–67 | Cancian* 1972 |
| 5) Mexico | 193 | Chiapas | 1966–67 | Cancian* 1972 |
| 6) Pakistan | 221 | Northwest Frontier | 1970 | Rochin* 1971 |
| 7) Pakistan | 350 | Punjab | 1970 | Lowdermilk* 1972 |
| 8) Philippines | 153 | Negros Oriental | 1966 | Pal and Polson* 1973 |
| 9) Philippines | 177 | Negros Oriental | 1952 | Pal and Polson* 1973 |
| 10) Taiwan | 159 | Taichung(hsien) | 1966 | Lionberger* and Chang 1968 |
| 11) Taiwan | 237 | Taichung(hsien) | 1966 | Lionberger* and Chang 1968 |
| 12) U.S. | 159 | Missouri | 1968 | Lionberger* and Copus 1972 |
| 13) U.S. | 167 | Missouri | 1968 | Lionberger* and Copus 1972 |
| 14) India | 676 | Three States [b] | 1967 | Roy et al. 1968 |
| 15) India | 1,145 | Andhra Pradesh | 1969–70 | Gartrell* 1974 |
| 16) U.S. | 316 | Wisconsin | 1962 | Wilkening* and Bharadwaj 1966 |
| 17) Japan | 91 | Ibaraki(prefecture) | 1955 | Lindstrom 1958 |
| 18) Mexico | 93 | Chiapas | 1965 | Cancian 1967 |
| 19) U.S. | 173 | Wisconsin | 1952 | Fliegel 1957 |
| 20) U.S. | 251 | Iowa | 1941 | Ryan and Gross 1943 |
| 21) U.S. | 252 | Kentucky | 1950 | Marsh and Coleman 1955 |
| 22) U.S. | 341 | North Carolina | 1949–50 | Wilkening 1952 |
| 23) U.S. | 423 | North Carolina | 1956 | Dean et al. 1958 |

NOTE: Data sets 2–14 and 16 have been cleaned and deposited in a public data bank. Current information on their availability may be had from Chief, Economic and Social Data Services Division, AID/PPC/PDPR, Department of State, Washington, D.C. 20523.

[a] A fuller list of references appears in Appendix D. An asterisk beside a name indicates that the data were provided by that individual. Randolph Barker provided the data for India 185; Everett Rogers for India 412; and Frederick Fliegel for India 676. Sources of data for the last 7 studies are recorded in Cancian 1967.

[b] Andhra Pradesh, Maharastra, and West Bengal.

These twenty-three data sets were assembled in three different ways. The first thirteen were gathered in the manner described in earlier sections of this chapter.* They form the core of this book. The next three were included because they are the empirical basis of three published comments (Gartrell, Wilkening, and Presser, 1973; Morrison, Kumar, Rogers, and Fliegel, 1976; and Gartrell, 1977) on my earlier work (1967, 1972) with the theory presented in Chapter 2. The final seven cases were the basis of my original paper (1967). These last two groups (the last ten cases), are carried through the next three chapters parallel to the core studies (the first group), but their analysis stays fairly close to that presented in the already published studies.

## Sampling

It is customary to ask what population the respondents represent and how adequately they represent it. That is: is the sample selected properly, and is it large enough to represent the population from which it was taken on the dimensions of variance that are of interest? With one possible exception that is noted, all the researchers whose studies are included here took standard precautions to insure that their respondents appropriately represent the populations studied. Many of them use community censuses—thus trading the sampling problem for the generalization problem. Note also that the sampling discussed here is that done *within* each study. As such it is relevant as a characteristic of the data set. It is not the sampling (of entire studies) relevant to the overall generalizations to be made on the basis of the present study.

Table 3 is a collection of short descriptions of sampling within each of the twenty-three studies. It also describes the relation of the sample I eventually selected for analysis to the sample studied by the original investigators. The diversity of procedure employed by the original investigators made standard tabular presentation of this information impractical. Some of the information in Table 3 is summarized in Table 4.

*The Mexico data sets are an exception insofar as they are based on my own field research, and were analyzed in Cancian 1972.

TABLE 3. *Detailed Sampling Procedures*

1) India 185

"All 185 farm households in Pedapulleru were included in the study and no further sampling procedure was necessary" (Parthasarathy 1975: 47).

2) India 412

702 respondents representing 85–90 percent of the household heads in 8 villages were interviewed (Roy et al. 1969: 74). 95 who owned no land were excluded by the original investigators. Respondents with no income or land on the independent variables used in this study were eliminated (reducing N to 568). People who worked off the farm 180 or more days annually were eliminated (reducing N to 412). The 8 villages were all within 10 or 12 miles of the city of Lucknow but were quite rural and isolated (p. 74). They were selected from 2 development blocks (see below); and attempts to divide them into modern and traditional "proved to be rather artificial" (p. 72). The authors consider them homogeneous for the purposes of analysis. "A 'block' is one of the basic units in development administration in India; it corresponds somewhat in land size to a county in other countries, but often has a population of about 100,000 people. A block usually contains about 100 villages" (p. 69n).

3) Kenya 540

540 farm households, 30 chosen at random from each of 18 villages, were interviewed. 25 percent of the respondents were women (Almy 1974: 256–260). 17 of the villages were geographically dispersed among the more than 70 villages in South Imenti Division (population 75,551, p. 29) of Meru District. The eighteenth, in North Imenti Division, was populated by people who moved there from South Imenti Division in the late 1950s (p. 23).

4) Mexico 108

142 of 147 household heads in Apas were interviewed (Cancian 1972: 165). From this sample of married men, those less than 25 years old and those still working cooperatively with their fathers were eliminated (N reduced to 119), and then those with no fields in the major study area (N = 11) were eliminated.

5) Mexico 193

232 of the 245 household heads in Nachig were interviewed (Cancian 1972: 166). From this sample of married men, those under 25 years old and those still working cooperatively with their fathers were eliminated (N reduced to 208), and then those with no fields in the major study area (N = 15) were eliminated.

6) Pakistan 221

226 farmers (Rochin 1971: 48) in 2 *thanas* (police jurisdictions) of Hazara District, Northwest Frontier Province, were interviewed. Lora thana is in Abbottabad *tehsil* (administrative district) and is about 100 miles from Oghi thana in Mansehra tehsil. The two thanas "hold a number of interesting contrasts and similarities" (p. 37) but are both within the rain-fed (*barani*) agriculture area (pp. 30–57). At least 45 villages are represented (p. 50),

TABLE 3. *Continued*

and the sampling was done under difficult conditions that made standard random sampling techniques impractical (p. 47). "The data from Lora and Oghi will be combined in order to develop a representative sample for Hazara District and the Himalayan barani" (p. 53). 5 respondents with 0 on the independent variable used in the present study were eliminated, making N = 221.

7) Pakistan 350

350 farmers representing a nonproportional stratified random sample of the 2,400 farmers with 2.5 acres or more in 30 villages of Khanewal tehsil, Multan District, Punjab Province, were interviewed. The 30 villages were randomly selected from 381 in the tehsil (Lowdermilk 1972: 74–85). The sample was stratified on farm size, and sampling fractions were used to "reconstitute" the sample for the analysis done below (see Chapter 4).

8) Philippines 153 and 9) Philippines 177

In 1952, 515 households chosen randomly from 23 of the 71 barrios in the Dumaguete trade area were interviewed. In 1966, 362 households, representing a stratified random sample of the 3,212 households censused in the 28 sample barrios of 92 then in the trade area, were interviewed (Pal and Polson 1973: 2–3). For 1952, respondents who were not primarily corn farmers were eliminated, yielding a sample of 177. For 1966, the data on corn farming was not available, and all those who were not primarily farmers were eliminated, yielding a sample of 153.

10) Taiwan 159 and 11) Taiwan 237

All farm operators in two villages (Shangfung N = 237 and Liupao N = 159) in Taya township of Taichung Hsien were interviewed (Lionberger and Chang 1968: 5–8, including note of the few operators not interviewed). Shangfung was chosen as an economically advantaged village and Liupao was chosen as an economically disadvantaged village (p. 6).

12) U.S. 159 and 13) U.S. 167

Farm operators in 2 communities, Ozark in poor, hilly southern Missouri, and Prairie in fertile, northwestern Missouri (Lionberger and Copus 1972: 76), were interviewed in 1968. A 1956 study with similar sample sizes (Lionberger and Chang 1965: 9) suggests that virtually all farmers in the communities were interviewed. Data available were for Ozark N = 221 and Prairie N = 173. Respondents working 200 or more days per year off the farm were eliminated, yielding samples of Ozark N = 167 and Prairie N = 159.

14) India 676

680 farmers with 2.5 acres or more, who were 50 years old or younger, were interviewed. They represented all the farmers who met the criteria in 8 villages spread over 3 development blocks, 1 in each of the 3 states listed in Table 2 (Roy et al. 1968: 9–10). 4 cases were lost in data cleanup and transfer.

*Continued*

TABLE 3. *Continued*

15) India 1,145

2,040 household heads sampled from 84 villages, 14 in each of 6 districts (1 district of high, middle, and low economic rank in each of 2 regions, Telangana and Andhra), were interviewed. "Within the 84 villages, heads-of-household were selected in proportionate random quota samples within three strata: cultivators, agricultural laborers and others" (Gartrell 1977: 321). The author uses all people who report some income from farming (N = 1,316). Here the additional stipulation that 51 percent of the household head's income must be derived from farming reduces the sample to 1,145.

16) U.S. 316

"A multi-stage area probability sample of unincorporated areas of Wisconsin was used. To be eligible for the sample, farm families had to meet all the following criteria: 1) married before January 1, 1961; 2) husband less than 65 years old; 3) husband worked as a farmer in 1961 either as his primary or secondary employment; 4) the couple grossed at least $1,000 from sales of farm produce in 1961. Of 495 qualifying cases, 102 farmers were not conducting a corn-dairy enterprise and 72 were not farming in 1952. They were excluded, leaving a final sample of 321. The reasons for these exclusions was to eliminate those cases for whom the measured innovations did not apply" (Gartrell et al. 1973: 394). Cases that had a 0 value on the independent variable used in the present study (N = 5) were also dropped.

17) Japan 91

Data were collected from 92 of 100 household heads in Tatsunokuchi *buraku* (neighborhood) in the community of Seki-mura. There was no information on the relevant variables for 1 respondent.

18) Mexico 93

"All married males, 27 years old and older, who were independent corn farmers and lived in the hamlet of Apas" are included in the sample (Cancian 1967: 920). Data were gathered from key informants in 1965. Data for the same community were gathered by more thorough interview procedures in 1967 (see Mexico 108 above).

19) U.S. 173

"A random sample of 200 Rock County, Wisconsin, farm owner-operators" (Fliegel 1957: 160). "The sample was restricted to farmers who had owned and operated their farms for at least three years; who had complete families, with children between the ages of 12 and 19 living at home; and who were employed off the farm less than 50 days in the year prior to the survey" (p. 160n). "No farm income-tax returns were available for 27 of the 200 sample operators" (p. 160)—reducing the sample to 173.

20) U.S. 251

"Practically every farm operator in two central Iowa communities" (Ryan and Gross 1950: 669) was interviewed. The communities were 15 miles apart in Greene County and were seen as basically similar by the investi-

TABLE 3. *Continued*

gators. Information on income was available for 251 farmers (Gross 1942: 96).

21) U.S. 252
All farmers in 13 neighborhoods selected on a judgment basis to represent all areas of Washington County (N = 393) were interviewed (Marsh and Coleman 1955: 292 and 1954: 1). The Cancian 1967 study dropped all but owners and part owners who grew tobacco as their principal farming activity—reducing the N to 252.

22) U.S. 341
"Complete enumeration was made of all male white farm operators (owners and share-tenants) in the localities selected who had obtained most of their gross cash income from farming the previous year" (Wilkening 1952: 8). The localities were 11 neighborhoods selected at random (with qualifications) from 241 identified in 3 noncontiguous counties (Harnett, Nash, and Wayne) in the central and upper coastal plain, North Carolina.

23) U.S. 423
"547 farm operators in eight counties of North Carolina were interviewed" (Dean et al. 1958: 121). The original investigators used up to 498 cases. Data for 480 cases was available for the Cancian 1967 study. 57 cases with less than 2 acres of corn seeded for grain were eliminated, yielding the sample of 423.

The diversity of sampling procedures also makes uniform appropriate language difficult to find. I have decided to use the word "sample" to refer to a group of individuals for whom there is data. While this usage is regrettably unconventional and somewhat misleading, all of the several alternatives I considered have worse faults. What the group (or sample) represents in terms of research procedure is described in Table 3 and the text immediately following.

I will begin with a specification of who is included in the final sample used for analysis and work outward. That is, I will begin with the individuals included in the N listed as part of the case identification, and work outward to the universe or population from which they are taken. These generalizations apply best to the first thirteen studies listed in Table 2.

People included for analysis typically are male household heads who farm a substantial part of the time. Female respondents were eliminated wherever possible; they were never a substantial proportion of the respondents interviewed by the original investigators (the Kenya study is an exception on both these counts). Original investigators usually interviewed household heads only. Part-time farmers were eliminated by cutting out respondents who worked off the farm 180 or more days per year—whenever off-farm work was measured. Respondents who reported that less than 51 percent of their income came from farming were cut in another case. In still other cases, the original investigators cut the sample to full-time farmers by some substantively similar definition. The few farmers who survived cuts like those just listed and still showed a zero value (not a missing value) for the independent variable (see Chapter 4) were also eliminated. In a few of the studies, farmers who concentrated on crops substantially different from those of the majority were eliminated. In sum, various efforts were made to insure that the sample used for analysis consisted solely of people who depended heavily on a reasonably homogeneous type of farming for their livelihood.

Perusal of Table 3 will uncover a number of unusual characteristics of samples or sampling procedures. The one that most attracts my attention is the requirement of 2.5 acres minimum land set for India 676 and Pakistan 350, in an effort to exclude farmers for whom technological innovation was impractical, impossible, or both. Other readers may find other characteristics relevant.

The relation of the "sample" to the "population" it represents and the "universe" from which it is taken is discussed in the following section.

### *Community of Reference*

Community of reference is of central importance in this study. The theory developed here emphasizes the importance of *relative* position in a community ranking system. The definition of the boundaries of the relevant community is especially important because, according to the theory, innovation both increases and decreases as rank increases. In order to operationalize these notions we must know the beginning and end of the relevant rank continuum; we must know the community of reference of the actor. Failure to identify the relevant community of reference may lead to aggregation that obscures curvilinear relation-

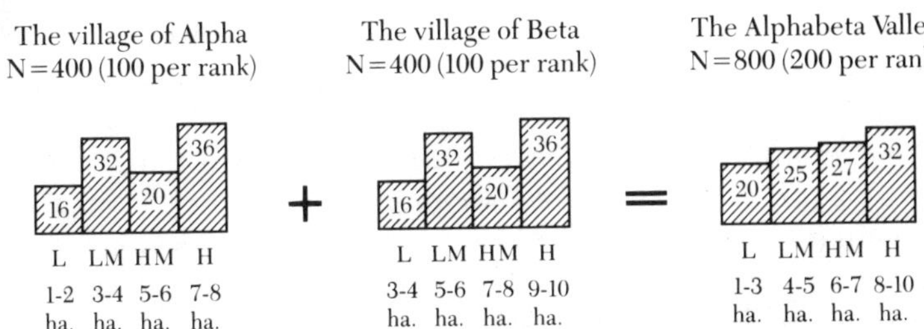


*Fig. 7. Potential effects of aggregation on curvilinear relationships. The number in each bar indicates the percent of farmers of that rank to adopt in a hypothetical Stage 1. The stage is defined as the first 26 percent to adopt. It is assumed that, within ranks within villages, differences in hectares cultivated imply no difference in innovation rate.*

ships. One way this might happen is shown in Figure 7.* The problem illustrated is not easily solved with routine precautions. It will be discussed briefly here and again in Chapter 6.

The community of reference is that group of alters within which the actor ranks himself in his day-to-day process of self-estimation or self-definition. This feature, and the use of the term "reference," makes the community of reference sound very similar to the complex reference-group idea (see Schmitt 1972). And my use of the term "reference group" in the sheet reproduced in Figure 6 adds to the connection in the present study. Nevertheless, "community of reference" as it is used here is different from the traditional reference-group idea in two fundamental ways. First, community of reference always indicates a membership group, never a group with which the actor compares himself or herself, or with which the actor *identifies* without belonging. Second, position within the group is crucial; the group is internally differentiated. By contrast, the reference group identified with or compared with is usually thought of as an undifferentiated type or group of individuals. On the other hand, the community-of-reference concept is very much like the reference-group concept in the important sense that it too requires delineation of the relevant set of alters.

The theory predicts that a low-middle-rank Iowa farmer and low-middle-rank Punjabi peasant will behave similarly relative

*I am indebted to A. K. Romney, who pointed out the relation of this problem to Simpson's reversal paradox (Bickel, Hammel, and O'Connell 1975; Gardner 1976).

to their respective high-middle-rank alters, but that Iowa and Punjabi farmers will not be important reference points for each other. It is not that Iowa and Punjabi farmers have never heard of one another, or that they are unconnected by the world market for agricultural products; it is rather that neither is a crucial immediate element in the other's day-to-day process of self-definition.

All this will come as no surprise to most readers. Some might even question whether the last few paragraphs improve much on the common-sense sociological statements of Merton and Rossi quoted in Chapter 1. Nevertheless, it is necessary to raise these issues because it is not easy to identify the appropriate group or society in the natural world; that is, it is not easy to effectively operationalize "community of reference" in a manner that avoids the problem illustrated in Figure 7. What follows offers some attempts at common-sense approximations. The question will be addressed again in Chapter 6, where we will be able to assess the effect of different definitions of the population studied on the tests of the principal hypotheses of the present study.

I expect that both spatial contiguity and total population size are relevant limits on community of reference—though neither is a necessary part of the notion of community of reference or the kindred reference-group notion. We know that class, caste, and ethnic differences cut across space and that modern communications vastly increase the potential number of relevant others; but, especially for the poorer people in any society, the community of reference will be in large part locally (spatially) defined and based heavily on face-to-face contacts. In contrast, it is likely that people at the top of any society will have relevant contacts outside any locally defined community. They are likely to belong to multiple spatially and culturally defined communities of reference. These complications, like others connected with the community of reference idea, are not subject to routine simplification.

In considering the relevance of spatial and population size characteristics to the study of individuals within a community of reference, it should be noted that a sample may adequately represent a community of reference in the sociological sense

TABLE 4. *Summary of Sampling Procedures and Universe Size*

| Study | Sampling procedure | Universe size[a] |
|---|---|---|
| 1) India 185 | Census of contiguous population | 1,332 |
| 2) India 412 | 8 villages in 2 contiguous blocks censused | 200,000 |
| 3) Kenya 540 | 1 division sampled | 79,995 |
| 4) Mexico 108 | Census of contiguous population | 778 |
| 5) Mexico 193 | Census of contiguous population | 1,418 |
| 6) Pakistan 221 | 2 thanas sampled | $10^{6+}$ |
| 7) Pakistan 350 | 1 tehsil sampled | 182,880 |
| 8) Philippines 153 | 1 trade area sampled | 63,332 |
| 9) Philippines 177 | 1 trade area sampled | 48,868 |
| 10) Taiwan 159 | Census of contiguous population | 1,145 |
| 11) Taiwan 237 | Census of contiguous population | 1,706 |
| 12) U.S. 159 | Census of contiguous population | 1,145 |
| 13) U.S. 167 | Census of contiguous population | 1,202 |
| 14) India 676 | 8 villages in 3 noncontiguous states | $10^{8+}$ |
| 15) India 1,145 | 1 state sampled | $10^{7+}$ |
| 16) U.S. 316 | 1 state sampled | $10^{6+}$ |
| 17) Japan 91 | Census of contiguous population | 655 |
| 18) Mexico 93 | Census of contiguous population | 670 |
| 19) U.S. 173 | 1 county sampled | 92,778 |
| 20) U.S. 251 | 2 communities in 1 county censused | 16,599 |
| 21) U.S. 252 | 13 neighborhoods in 1 county censused | 12,777 |
| 22) U.S. 341 | 8 neighborhoods in 3 noncontiguous counties censused | 343,582 |
| 23) U.S. 423 | 8 counties sampled | $10^{6+}$ |

[a] Universe size for cases involving census of contiguous populations are uniformly "estimated" with this formula: the number of farmers, plus 20 percent of that number (to compensate for other households in the community), times 6 (persons per household); or, simply, universe size = 7.2N. This produces rough estimates for comparison with figures for studies that used other sampling procedures. Estimates for India 412, Kenya 540, Pakistan 350, and Philippines 153 and 177 are based entirely on information from Table 3. Hazara District (Pakistan 221) had a population of 1.6 million at the time of study (Rochin 1971: 33). The following estimates are based on census figures. The 1971 population of just the 3 states in India 676 exceeded 137,000,000. Andhra Pradesh itself (for India 1,145) was 50,412,235. Wisconsin in 1960 (for U.S. 316) was 3,951,777. Rock County, Wisconsin, in 1960 (for U.S. 173) was 92,778. Greene County, Iowa, in 1940 (for U.S. 251) was 16,599. Washington County, Kentucky, in 1950 (for U.S. 252) was 12,777. For U.S. 341 I took the 1950 population of Harnett, Wayne, and Nash counties, nearby noncontiguous counties in North Carolina, and doubled it. North Carolina in 1960 (for U.S. 423) was 4,556,155. In most cases rural populations are smaller than total populations.

even if it is not representative of the geographical extension of the actual community of reference. For example, a single village may be an adequate representation of a community of reference that actually includes a number of similarly structured villages. Likewise, a random sample of a state made up of numerous very similar separate communities of reference might adequately represent any one of them. On the other hand, a single village cannot represent a diverse set of villages, and a random sample of a sociologically diverse state cannot represent each locally defined community of reference.

Table 4 displays a summary of sampling procedures and an estimate of the size of the universe sampled for each of the twenty-three studies assembled here. In the table the word census is used to describe a data-gathering procedure that covered all or almost all the farmers in a contiguous area. The universe for which number of inhabitants is estimated is defined as the smallest ungerrymandered contiguous area that includes all the individuals in the sample. Estimates of population size where the sample is not a census of a contiguous area are rough, but they seem accurate enough given the very substantial range covered. Notes to the table explain the estimating procedures.

Speculation about the differences shown in Table 4 will be deferred until after description of the variables has been completed in Chapters 4 and 5.

### *Summary*

Twenty-three studies done by rural sociologists, anthropologists, and agricultural economists were assembled for secondary analysis. Seven came from a previous study (Cancian 1967), three came from published comments on that study, and thirteen were newly assembled for the present study. In most cases the original investigators provided their raw data. With a few exceptions the original studies were censuses or systematic samples of farmers in a single community, trade area, or county. The theory being tested requires that all the farmers in each study be in a single community of reference. Initial efforts to specify the community-of-reference idea and to identify appropriate communities of reference in the assembled data were made.

*Chapter Four*

# *Economic Rank: The Independent Variable*

A farmer's economic rank is his place in a hierarchy that includes all the farmers in his community. This apparent unidimensional simplicity becomes complicated for a variety of reasons that will be explored in this chapter. In the last analysis, the complication stems from the fact that economic position is important to people, and most people expend great effort gaining and maintaining their economic rank.

The chapter is divided into four parts. First, the general theoretical and conceptual issues about economic rank will be discussed. This focus continues the concern with general issues in social stratification with which Chapter 3 ended. Then the chapter turns to the research tradition from which the data are borrowed and its implications for choice of measures of economic rank in farming communities. Third, the size and distribution of economic resources in the communities studied is reviewed; and, finally, the mundane and the general issues involved in the manipulation of data on economic rank are discussed.

## *What Is Economic Rank?*

Rank is the individual's place relative to other people in the hierarchy of a community. The definition of the limits of the community and the problems associated with establishing them were discussed in Chapter 3. Here the focus is on the nature of rank within that community.

It is easy to see that stratification or inequality has both relative and absolute aspects. The same $100 per week income that places a family near the bottom of an American urban community would make most third-world agricultural families virtual princes in their villages. While this contrast across stratification systems is commonly understood, the absolute and relative aspects of stratification within systems are more confusing.

Three basic notions of inequality are useful in sorting out these confusions. They are based on: 1) an absolute scale on the underlying variable used to establish rank; 2) a share in the distribution of the community's total accumulation of the same resource; and 3) a count of the people who have a larger (or smaller) share than ego. Miller and Roby (1970) provide an excellent discussion of the subtleties of definition involved in the first two notions listed here, and ample evidence for the political and social importance of the different definitions. What follows is meant only to firmly establish the contrast of the first two with the third notion.

Absolute level on an underlying resource variable (for example, dollars of income) does not involve the individual's relation to other people; it is measured without regard to the amount other people in the community may have. Such a notion is behind the most common efforts to reduce or eliminate poverty. Setting the poverty level at a given number of dollars makes it possible to change "poor" into "nonpoor," whatever the level of income of other people in the society. Of course, it is now commonly understood in the United States that the definition of essentials characterizing the nonpoor may change—to include, for example, a color television set. That is, it is understood that the definition of essentials involves relative elements. Nonetheless, the nonrelative definition continues to temporarily serve the political process, especially when inflation virtually insures statistical movement out of poverty that is defined by an absolute dollar level.

Before going on to the second notion, structural mobility should be mentioned. Structural mobility represents a phenomenon that is somewhat like the statistical escape from poverty described above. Structural mobility occurs when, for example, the number of skilled jobs in the economy increases as a proportion of the total jobs, and some workers move up from unskilled to skilled jobs. A moment's thought will make it clear that they may do this without change in their relative position in the job hierarchy or change in the share of the total wealth or income they control. In fact, their absolute earning level may not increase, either—though it is commonly thought to do so in response to their greater productivity as skilled workers. Historically, structural mobility has often materialized as an increase in position relative to a culturally defined scale established in the previous generation. As such it provided much psychologically real upward mobility that involved little change in rank or increased share in the community's resources. In the present context, the notion of structural mobility lies between the first and the other notions in my list of three. In the form of structural mobility that is of most interest for expository purposes, the market basket is bigger for the upwardly mobile person, but rank and share are not. This form of structural mobility is thus characterized by the presence of absolute mobility and the simultaneous absence of relative mobility.

The second of my three notions concerns the share an individual (or class of individuals) has in the distribution of the community's total accumulation. It has a lot to do with the individual's (or the group's) relation to other people. As Miller and Roby (1970: 22) point out, this notion frankly recognizes inequality within the community rather than absolute wealth level as the crucial idea. In fact, Miller and Roby properly (for their purposes) divide this notion into two parts. The first one defines the poverty level as a fixed fraction of the median or mean income in the overall community. The second looks at the proportion of the total income that goes to, say, the bottom 20 percent of the population. Clearly, the various possible implementations of these definitions have different implications, depending on the exact shape of the resource distribution in the community studied.

For my purposes these two variants of the share notion have two things in common. First, they are both relational, that is, they both require a definition of a community of reference in terms of which the distribution they describe can be viewed. Second, they both permit calculation of degrees of inequality within the community. There is more inequality when 30 percent of the population has incomes of less than one-half the median income than when 20 percent has such incomes. There is more inequality when the bottom 20 percent receives 5 percent of the total income than when it receives 7 percent. All these calculations of degrees of inequality are based on the absolute (ratio) scale of the underlying variable—in this case, income. In sum, to be properly defined, share-of-resources notions require information about both the community of reference and the absolute level of resources controlled by each part of the community.

The third notion requires information about the community of reference but not about the absolute level of resources controlled. It is the well-known notion of hierarchy based on relative rank. This is the basic notion underlying much of the thinking in the present study. I have placed this statement at the end of such a long discussion of other notions of economic rank (or rank in general) because of the confusion that often surrounds it in practical research situations. This confusion goes on while the idea of hierarchy has a firm and clear place in abstract theoretical discussions.

Some researchers see rank measures (measures with no quantifiable underlying variable) as a necessary compromise in situations where measurement is difficult or costly. In effect, they see interval or ratio scales as always more desirable than ordinal scales. Although it is true that interval and ratio scales can always be converted to ordinal scales, while the opposite is not true, this characteristic of measurement and statistical manipulation must not be confused with the world of sociological phenomena to which it is often related.

Human relationships need not be characterizable by an independently measurable ratio scale. In fact, it is my concern here to emphasize the importance of a difference in rank, no matter what the "size" of the difference between ranks on any underlying variable. Higher and lower, when they can be clearly determined and are fairly stable, function without large absolute differences. The important thing in an interaction may be who has to defer to whom, not the fact that one person is 10 percent or 100 percent poorer (or richer) than the other. The important thing in a community may be how many people are higher and lower than the actor, not how much higher or lower they are on an absolute scale. In this book, rank in the relative sense is most often the concept underlying the thinking about stratification within the community.

Nevertheless, the absolute level of the underlying variable often has significance that cannot be denied if natural situations are to be understood. For example, absolute farm size may often be relevant to what a farmer does. At the extreme, a huge tractor cannot turn around in a tiny vegetable plot. More commonly, "expensive" equipment cannot repay its cost on many "small" farms. In the theory stated in Chapter 2 these realities are incorporated into the facilitating effect of rank. Everywhere, high rank means more control of various resources on an absolute scale. So, relatively high rank means a relatively high probability of being above financial, technological, "informational," and other absolute thresholds. While I now want to emphasize the relative nature of economic rank, the constraints embodied in the absolute size of land, or other resources often used in measures of economic rank, need to be clearly recognized in any attempt to understand behavior in natural settings.

Finally, I believe that the confusion of absolute and relative measures of rank within systems stems in part from the practical interchangeability of measures in many situations. Subtleties aside, people, both social scientists and farmers, are apt to talk in terms of one measure when they mean (for the most part) the other. Since it often makes no difference to communication, it is not noticed. That is, we may say "She's worth half a million," when we mean she's rich relative to others. And, we may say "He's rich" when asked whether he can pay $15,000 for a new car. Both our tendency to slip effortlessly back and forth (even when it is not appropriate), and the tendency of farmers the world over to do the same thing, must be kept in mind as we deal with customary measures of rank and absolute units of resources.

### Measures of Economic Rank in Farming Communities

In practice, the choice of measures used here is limited by the measures used in the research tradition from which the data are borrowed. Fortunately, this tradition has a fairly consistent and concrete set of practices. The most common appropriate measures for the resource rank or economic rank variable are: area of land owned or farmed, value of farm products produced or sold, and net income from farming. They will be discussed in turn.

Area of land farmed is the most widely available measure that can be used to index economic rank. It is often called size of operation. Measures of land area, of course, suffer to the degree that the land is not of consistent quality within a community. A hectare of irrigated garden is not equal to a hectare of range. Thus, size of operation as a measure of economic rank is less appropriate the more diverse are the lands worked by the population studied.

In many communities a substantial number of farmers rent or sharecrop a substantial part of their holdings. In many places farmers rent out some of the land they own and rent from someone else part of the land they farm (that is, they rent out and rent in). Although area owned may be a better measure of wealth in the few cases where there is a great difference between land owned and land operated, in my judgment area operated is the better overall index of economic rank.

Tenancy is, of course, an important factor in the sociology of farmer behavior. Accordingly, for a small number of cases I established "owners' samples," from which tenants and farmers who were mostly tenants were eliminated, and retested the principal hypotheses. None of the results were substantially different from "full samples" that included tenants. On the basis of this cursory investigation the study of tenancy was abandoned in favor of other research priorities.

The value of farm products produced or sold might also be used to index economic rank. However, it is not measured as frequently as is land area, and it has qualities that make it more complex than land area. In peasant communities the value of product sold may be a poor measure of the value of production because of the substantial share of production that goes directly to family consumption, but this feature need not interfere with its use for internal ranking of farmers. In a mixed farming situation in more industrialized agriculture, however, value of product or gross farm income figures can be misleading for comparison of farmers in different kinds of farming. A large part of a feed lot's "product" sold may be "circulating capital" rather than "value added," while that is not true for a dairy farmer. Thus, value of product measures, while they eliminate some of the problems of land area measures, have parallel problems of their own.

Net income from farming is an excellent index of economic rank which eliminates many of the problems of underlying inconsistencies that may apply with either land area or value of produce (gross farm income). Since farmers are less likely than salaried or wage workers to know their incomes, and since they are not known to be more relaxed about revealing them, net farm income is hard to measure. Thus, it is seldom measured.

The distinction between wealth and income could have important implications for the development of the theory stated in Chapter 2. Wealth, insofar as it embodies the long-term potential to produce an income stream, is more like the resources discussed in the theory than actual (present time) income is. Thus, if in a given case there were a significant difference between a farmer's income rank and his wealth rank, the choice between those variables would be important. Since the difference would be difficult to operationalize for a substantial proportion of the communities studied here, and since I have little evidence that whatever difference there might be between wealth rank and income rank could be picked up by the relatively gross measures used here, the distinction is ignored in what follows.

Finally, the level of living index, which is included in many studies that follow the rural sociology research tradition, needs to be considered briefly. Since level of living indexes typically include modern or innovative consumption items, they may correlate, for reasons of the individual's personal style, with the indexes of innovation in production used in the present study. Thus the index of economic rank provided by the level of living index is confounding given the goals of the present study. For this reason, it is not used below.

Before leaving this general discussion of the practical measures available for the independent variable, I should note that all three of the measures seriously considered (land area, value

TABLE 5. *The Independent Variables in Each Study*

| Study | Variables |
|---|---|
| 1) India 185 | Farm size |
| 2) India 412 | Value of products raised; Land farmed |
| 3) Kenya 540 | Composite wealth scale; Land owned |
| 4) Mexico 108 | Land farmed |
| 5) Mexico 193 | Land farmed |
| 6) Pakistan 221 | Wheat harvested |
| 7) Pakistan 350 | Wheat seeded |
| 8) Philippines 153 | Value of products sold; Cropland |
| 9) Philippines 177 | Value of products sold; Cropland |
| 10) Taiwan 159 | Land operated |
| 11) Taiwan 237 | Land operated |
| 12) U.S. 159 | Net farm income; Land operated |
| 13) U.S. 167 | Net farm income; Land operated |
| 14) India 676 | Value of product raised; Land cultivated |
| 15) India 1,145 | Value of product raised; Land cultivated |
| 16) U.S. 316 | Gross farm income |
| 17) Japan 91 | Farm size |
| 18) Mexico 93 | Land farmed |
| 19) U.S. 173 | Net farm income |
| 20) U.S. 251 | Net income |
| 21) U.S. 252 | Land operated |
| 22) U.S. 341 | Cropland operated |
| 23) U.S. 423 | Land in crops and improved pasture |

of produce, and net income) are usually measured on ratio scales. Thus, at this point at least, it is possible to carry forward both the ordinal (relative) measure of rank and the ratio (absolute) measure of the underlying variable.

The variables used to index economic rank in this study are listed in Table 5. For those among the top sixteen cases where more than one variable is listed, the first listed is judged to be the best measure. In most such cases an income or value of product measure is preferred to land area.

## *The Size and Distribution of Economic Resources*

Both the absolute size of economic resources and their distribution in the community are important to understanding the

TABLE 6. *Size and Distribution of Farmlands (in hectares)*

| Study | Full farm[a] | Mean | 5%[b] | 25% | Median | 75% | 95% | Gini |
|---|---|---|---|---|---|---|---|---|
| 1) India 185 | X | 4.7 | .4 | 1.3 | 2.4 | 5.9 | 16.2 | .55 |
| 2) India 412 | X | 1.3 | .2 | .7 | .9 | 1.6 | 3.5 | .39 |
| 3) Kenya 540[c] | — | — | .4 | 1.2 | 2.4 | 6.1 | 40.1 | — |
| 4) Mexico 108 | X | 2.2 | .8 | 1.5 | 2.3 | 2.3 | 4.5 | .29 |
| 5) Mexico 193 | X | 2.6 | .8 | 1.5 | 2.3 | 3.8 | 6.0 | .32 |
| 6) Pakistan 221[c] | — | — | .1 | .3 | .5 | 1.0 | 2.0 | — |
| 7) Pakistan 350[c] | — | — | —[d] | —[d] | —[d] | —[d] | —[d] | — |
| 8) Philippines 153 | X | 1.7 | .2 | .7 | 1.1 | 1.8 | 5.0 | .48 |
| 9) Philippines 177 | X | 1.6 | .2 | .6 | 1.0 | 2.0 | 4.0 | .46 |
| 10) Taiwan 159[e] | X | 1.0 | .1 | .4 | .7 | 1.2 | 2.9 | .45 |
| 11) Taiwan 237[e] | X | 1.0 | 0.0 | .5 | .8 | 1.3 | 2.8 | .36 |
| 12) U.S. 159 | X | 191.6 | 32.4 | 88.3 | 138.9 | 232.1 | 496.5 | .41 |
| 13) U.S. 167 | X | 132.4 | 29.2 | 49.0 | 85.1 | 168.1 | 405.0 | .45 |
| 14) India 676[c] | — | — | 1.1 | 1.6 | 2.7 | 5.4 | 12.1 | — |
| 15) India 1,145[c] | X | — | .4 | 1.6 | 3.0 | 5.4 | 15.0 | — |
| 16) U.S. 316[e] | X | 86.0 | 7.5 | 50.0 | 70.0 | 95.0 | 160.0 | —[f] |

NOTE: All data originally in acres was converted at 1 acre to .4047 hectares. The cases where neither hectares nor acres were used in the original data were converted to hectares as follows: India 412, 4.3 *bighas* = 1 hectare; Mexico 108 and Mexico 193, 1 *almud* = .75 hectares; and Pakistan 221, 1 *kanal* = .05 hectares.

[a] X means the variable includes a reasonable approximation of full farm size.

[b] Figure in the table is farm size for farmer(s) at the fifth percentile rank, etc.

[c] Kenya 540 gives a land owned measure, but Almy states that land owned is not a good measure of economic rank (1974: 237). Pakistan 221 includes only wheat harvested in a system where maize and wheat are both important (Rochin 1970: 21). Pakistan 350 and India 676 limit the sample to farmers with 2.5 acres or more. Sampling for India 1,145 slightly overrepresents the rich (Gartrell 1977: 321). Means and Gini coefficients are not included for these 5 cases. Median and other percentile points are included in the table but should be interpreted in the light of the distortions noted here.

[d] Nonproportional stratified random sample. Reconstructed for original rank distribution (see Appendix A) but not for area variable.

[e] Roughly interpolated from grouped data.

[f] Gini not calculated.

stratification system and its relation to adoption of new agricultural practices. These features of the measures used in this study are displayed in Tables 6 and 7. These tables include only the first sixteen studies (the first and second of the three groups). The final seven studies (the third group) are not included here. They were analyzed in Cancian (1967), and details on the distributions are not available.

On an absolute scale most of the farms studied here are small

TABLE 7. *Size and Distribution of Income Variables (in local currencies)*

| Study | Full income[a] | Mean | 5% | 25% | Median | 75% | 95% | Gini |
|---|---|---|---|---|---|---|---|---|
| 2) India 412, rupees[b] | X | 350 | 40 | 110 | 220 | 440 | 1,080 | .51 |
| 3) Kenya 540[c] | — | — | — | — | — | — | — | — |
| 8) Philippines 153,—pesos[d] | X | 361 | 0 | 56 | 180 | 413 | 1,200 | .63[e] |
| 9) Philippines 177,—pesos[f] | X | 110 | 1 | 1 | 18 | 150 | 410 | .76[e] |
| 12) U.S. 159, dollars | XX | 7,531 | 1,000 | 3,500 | 6,500 | 10,000 | 20,000 | .40 |
| 13) U.S. 167, dollars | XX | 2,050 | 0 | 622 | 1,500 | 2,700 | 5,800 | .50 |
| 14) India 676, rupees[g] | X | 4,319 | 373 | 942 | 2,256 | 4,555 | 13,966 | —[h] |
| 15) India 1,145, rupees[i] | X | 5,834 | 400 | 1,348 | 3,000 | 6,360 | 20,545 | —[h] |
| 16) U.S. 316, dollars[j] | X | 9,200 | 3,333 | 6,200 | 9,900 | 14,500 | 23,750 | —[k] |

[a] X indicates a "gross" measure, e.g., value of product produced or gross farm income. XX indicates net farm income.

[b] A dollar equaled 4.8 rupees in 1963, the time of the fieldwork (Pick 1968).

[c] Kenya 540 has no income variable, but it does have a wealth index that is the preferred variable (Table 5).

[d] A dollar equaled 3.9 pesos in 1966, the year of the fieldwork (Pick 1976).

[e] These rows show only the value of products sold, and exclude the value of products consumed, and the Gini coefficients are correspondingly distorted.

[f] A dollar equaled 2.34 pesos in 1952, the year of the fieldwork (Pick and Sedillot 1971).

[g] A dollar equaled 7.5 rupees in 1967, the time of the fieldwork (Pick 1976, confirmed by E. Rogers).

[h] Gini coefficient not calculated because of nonrepresentative sampling (see notes to Table 6).

[i] A dollar equaled 7.5 rupees in 1970, the time of the fieldwork (Pick 1976).

[j] Roughly interpolated from grouped data.

[k] Gini not calculated.

compared to what students of agriculture in the United States have come to expect. In fact, they may be too small for adoption of many practices that change agents would like to spread. In 1960, Lionberger concluded on the basis of data from Missouri that "a critical minimum size of about 140 acres [56.7 hectares] for the adoption of the farm practices considered was therefore suggested" for that situation. He went on: "Perhaps a minimum size of farming operation necessary for successful adoption can be specified for most farm practices and most areas" (1960:101). Now, almost twenty years later, after the internationalization of research on adoption of new agricultural practices, it is clear that, in comparative perspective, 140 acres is a large farm indeed. Many new farm practices can be used in many areas with much smaller farm sizes.

The data sets displayed in Table 6 have no precisely known quantitative relation to the universe of all farming communities in the world, and thus no conclusion about typical farm size can be drawn. It is true, however, that there are a great number of communities like the ones studied here. Thus it is likely that a conclusion made on the basis of this sample of communities applies to a substantial part of the world's population of farming communities.

The absolute size of farms described in Table 6 is in each case limited by the rules (see Chapter 3) applied to eliminate part-time farmers who had primary commitment to a nonfarming source of income. Thus, while the samples include many farmers who have other sources of income, they also exclude many rural people for whom farming may be an important auxiliary source of income. In two cases (India 676 and Pakistan 350) the original sampling (see Table 3) was restricted to farms of 2.5 acres or more.

The income figures are shown in Table 7 along with conversion rates prevalent at the time of the original studies. Note that in many cases the figures shown in the table are not net total income figures.

The distribution of land and income within the communities shown in Tables 6 and 7 is positively skewed. That is, there are relatively many small farmers and relatively few large farmers in each community. This comes as no surprise to any student of stratification or everyday life.

The Gini coefficient, shown in the righthand column of Tables 6 and 7 for the cases in which it is appropriate to calculate it, is a summary measure of inequality that goes from 0 (equality) to 1 (total concentration of resources in the community).

### *Manipulation of the Data*

Manipulation of the measures of the independent variable is crucial in the present study. To test the hypotheses, the predominantly ratio scales used by original investigators will be transformed into ordinal scales—into rank orders. This basic transformation of the data will be described in concrete terms in this section. Its implications pervade the study.

The distributions of the independent variable in the various samples are similar in being positively skewed, but they are otherwise quite different—in both their absolute size and in the degree of inequality they show. These two characteristics of the data are temporarily lost while we focus on the variables in their rank form. Some details of effects of this transformation to "ordinal analysis" are discussed in Appendix B.

It is important to note, however, that the "loss" of information on absolute size and shape and range of the distributions is replaced by the "gain" of a relational measure (rank) that is comparable across studies. As noted at the beginning of this chapter, the rank measure has important sociological characteristics that a measure based on a ratio scale can have only fortuitously. That is, if the distance between individuals in a rank-ordered population is sociologically equivalent for all adjacent pairs, then perfect isomorphism with an underlying wealth scale (such that the "dollar" difference between each pair is also equivalent) is possible, unlikely, and in any case irrelevant to sociological analysis.

The theory stated in Chapter 2 demands that at least four ranks be specified. In this study two divisions into four ranks are used. The first, and simplest, way to divide the rank continuum into four ranks is to use equal ranks or quartiles to represent the low, low middle, high middle, and high ranks specified in the theory. A second procedure in which the ranks are represented by 30 percent, 30 percent, 20 percent, and 20 percent of the continuum respectively is also explained below.

The quartiling procedure was used in my initial research (Cancian 1967). It has one lasting advantage and some serious defects. Its advantage is that it makes small expected cell frequencies as large as possible. That is, by dividing the rank continuum evenly it provides comparatively large cell sizes for the smallest cell even with a small overall sample size. Any uneven division of the continuum reduces the size, and thus the statistical stability, of at least one of the four ranks. When working with sample sizes under two hundred, even division has much to recommend it. The specific rules used to divide the rank data into quartiles and other divisions discussed below are described in Appendix A.

On the other hand, the quartiling procedure has a number of defects. Intuitively, and in terms of the absolute quantities involved, the positively skewed distribution of the underlying variable suggests that the lower ranks, conceived in terms of sociologically significant groups, may be larger and the higher ranks smaller in numbers of farmers. It is certainly true that there is less absolute difference between the low middle and low quartiles than between the high middle and the low middle quartiles (see Table 6). On these grounds, and on the general grounds that agricultural populations have "triangular" rank structures, Morrison et al. (1976) have argued that, if the dip exists, it should be expected to come at a point higher up the rank continuum. With this in mind they make four ranks which have roughly the following ratio from low to high: 6 : 3 : 2 : 1. Even with their large sample size (N = 557), their highest rank includes only forty-five farmers. Given the appropriateness of their argument, and the countervailing realities of the sample sizes in the assembled data sets, I have adopted a procedure for cutting the rank variable in a 3 : 3 : 2 : 2 ratio as well as in the 1 : 1 : 1 : 1 ratio implied by the quartiles (Q). This "30/20" (T for thirty is used in table labels) division goes in the direction suggested by Morrison et al. without drastically reducing the number of farmers in the higher ranks.

The theory, of course, does not specify the relative size of the four ranks. Nor does it explicitly state that they will be the same relative size in every case. Given the extraordinary diversity of locations and situations from which the data were taken, it would not be surprising if both size and location of the different ranks (taken now as ideal types) varied from community to community. Thus the best procedure for testing the theory would not be influenced by the fortuitous and irrelevant combination of cutting point and naturally-occurring distribution.

The theory does not even specify four ranks; it specifies a minimum of four ranks. The division into four ranks is a direct product of efforts to develop practical tests for the hypotheses. In that context typological thinking encouraged division into discrete ranks, but probabilistic thinking clearly suggests that the empirical relationship is most apt to be the continuous sort of thing illustrated in Figure 1.

Continuous measures would avoid problems mentioned above and allow the shape of the relation of rank and innovation to be determined inductively in terms appropriate to the particular community studied. Several efforts to use continuous measures to inductively assess the shape of the relationship between rank and innovation are described in Appendix B. They include use of deciles, percentiles, and moving averages rather than quartiles, and the exploration of regression analysis with both absolute and rank variables.

### Summary

In sum, economic rank is a special kind of measure of stratification. It is based on relative position of individuals in the group, not on any of the other frameworks customarily used in the analysis of stratification. As such, economic rank is a sociological, not an economic, variable. It is nonetheless true that the absolute scale and the material constraints that usually go along with economic rank cannot be forgotten. Thus the rank variable quickly involves much more than a single simple hierarchy. The concrete measures used to index economic rank are area of the land farmed, and in some cases income measures of various sorts. These measures transformed into ranks provide a basis for comparison across the diverse populations included in this study, despite the fact that the absolute size of farms and the distribution of land resources within communities varies from population to population. The rank continuum will be divided in a variety of ways and by a variety of methods, because the theory presented in Chapter 2 does not specify where in the continuum the predicted dip will be.

*Chapter Five*

# *Innovation: The Dependent Variable*

The concept of innovation, like the concept of economic rank discussed in Chapter 4, is richly complex and subject to many alternative interpretations. Here I will follow a procedure parallel to that used in Chapter 4. First the concept of innovation and its place in the overall theory will be discussed. Then I will review the measures of innovation usually used in studies of agricultural innovation, then the farming situations and accompanying measures included in the particular studies used here; and, finally, I will describe the specific manipulations of the data used in testing the theory.

## *What Is Innovation?*

Innovation has many meanings. Since my purpose here is to use the term in a limited sense, I will not try to cover the range or even the core of its meanings. The term means here as much as the reader is willing to allow it to mean given the specific research operations used to give it concrete meaning. In what follows I will try to explain what I imagine it to mean—that is, what I hope the measures measure.

An innovation is new and different. As such, innovation is a relative term, for when it refers to a practice, a technique, or an object, it is appropriate only as long as the practice, technique, or object is new and different. When the practice is old, when it is commonly expected, when it is the standard thing it is no longer an innovation. An innovator is a person who adopts the practice, technique, or object while it is still an innovation.

In our culture innovation is good, and those who try to be

new and different are generally thought to be praiseworthy. This "moral" aspect of innovation is not crucial to my efforts here and will not be pursued further. However, it is worth noting that people often hesitate to apply the name innovation to a practice, technique, or object that "fails."

Technological innovations usually spread from the political and cultural center to the periphery of influence and culture. They are often involved in the relations of subordination and superordination between the periphery and the center because, for many innovations, the center controls the supply of material inputs. In the present study I have nothing further to say about this feature of innovations.

It is the unknown quality of an innovation that makes adoption or trial risky or uncertain. That is, the farmer using the new practice does not know what his potential for profit will be. He hopes it is better than that of alternative practices, but, given the nature of the situation, he cannot be sure. As a new practice becomes known, the uncertainty gives way to familiarity. The farmer knows what to expect from the practice, which is by then no longer new—no longer risky because of unfamiliarity.

Of course, substantial risk due principally to climate is seldom absent from agricultural production. Market fluctuations introduce still other substantial risks for commercial farmers. These background sources of risk are present for all farmers in a community and do not distinguish the innovators from the rest of the population. While I have tried elsewhere to distinguish this kind of risk from the uncertainty associated with newness (1972, 1979b), the distinction is not crucial here. For now, we need only keep clear the idea that some uncertainty or risk is associated with newness.

This aspect of an innovative practice, technique, or object connects it to the uncertainty that is crucial to the theory stated in Chapter 2. Of course, this interpretation of what happens when a farmer tries or adopts a relatively new agricultural practice is a hypothesis. It makes sense in general, given what is known about the concrete situation of farmers, but whether or not it is to be taken as a useful conceptualization of the situation depends, in part, on the results of this study.

In contrast, the fact that an innovation is unknown as well as new and different is normally not emphasized by the "diffusion of innovations" research tradition. This is because, in large part, this research tradition is harnessed to the effort to promote innovations through the increase of information and certainty about the implications of their use. That is, the diffusion of innovations research tradition is an integral part of an effort to promote certainty. None the less, the measures used by researchers in that tradition to record the diffusion of innovations fit well with the conceptualization in terms of unfamiliarity and uncertainty used here.

### *Measures of Innovation in Farming Communities*

The innovation variable described in Chapter 2 defines the innovator as the person who adopts a new practice *before* his or her fellows. The practice involved should be an economically important one such that the decision to try it might substantially affect the actor's economic rank in the community. Thus, by using the new practice before its implications are fairly certain, the innovator may make a killing or suffer a loss relative to his or her fellows.

The present study of farmers' practical choices is limited by the measures used in the research tradition from which the data are borrowed. The general notions discussed above must be operationalized with the available measures. As was noted above, this tradition has a fairly consistent and concrete set of research practices which provide the kinds of measures needed to operationalize the theory.

The prototypical measure of innovation was made famous by the Ryan and Gross (1943) study of the spread of hybrid corn seed in Iowa (here U.S. 251). It is year of first use of an important practice. In the area studied, hybrid corn seed adoption began in 1927 and was virtually complete by 1941. As is shown in Table 8 most of the farmers adopted in the middle years of that period. The spread of high-yielding variety (HYV) wheat and rice in Asia in recent years offers similar opportunities for study of a single important change. This type of measure of the innovation variable conforms closely to the theoretical notion set out in Chapter 2.

Indexes based on the number of new practices employed by

TABLE 8. *Adoption of Hybrid Corn Seed in Two Iowa Communities*

| Time period | Number adopting |
|---|---|
| 1927–33 | 23 |
| 1934–36 | 72 |
| 1937–39 | 145 |
| 1940–41 | 17 |
| TOTAL[a] | 257 |

SOURCE: Ryan and Gross 1950: 678–79.
[a] Of 259 operators studied, 2 had not yet adopted in 1941 (Ryan and Gross, 1943: 19).

a farmer at a given point in time (usually the time of the study) are also common in the literature on agriculture innovation. Table 9 lists the ten innovations used for the index in the India 676 study, along with the percentage of all farmers who had tried each innovation at the time of the study. On this index a farmer's score could range from zero to ten. These indexes have the advantage that they avoid the need to ask the respondent to recall an event from the past; and they have the additional advantage that they may be used where a complex of new practices (rather than a single important one) is involved. They have corresponding disadvantages, however. First, to make them parallel to the year of first use measure, the indexes require the assumption that the use of more items at a given point in time implies the use of at least some of them at an earlier point than others in the community. This assumption is easily made on a common-sense basis, and I have never found any reason to question it, given the essentially crude level of measurement employed in studies of this kind. The second disadvantage of these indexes is that, as they are usually used, they require the assumption that all items are of equal weight, that is, have equal uncertainty and equal cost. Since it is unlikely that this is ever exactly true, the conscientiousness of the original investigator must be depended upon for elimination of grossly inappropriate items.

In mixed farming situations indexes involve an additional complication. Farmers in the same community of reference may be working with different crops that require different technol-

TABLE 9. *Innovations Used for Trial Index in India 676 Study*

| Innovations | Percent of respondents who tried | Innovations | Percent of respondents who tried |
|---|---|---|---|
| 1. Ammonium sulphate | 72% | 7. Green manure | 40% |
| 2. Animal inoculation | 68 | 8. Cultivator | 13 |
| 3. Fertilizer mixtures | 60 | 9. Improved breeding of cattle | 11 |
| 4. Rat poison | 59 | 10. HYV | 9 |
| 5. Insecticides | 55 | | |
| 6. Superphosphate | 50 | | |

SOURCE: Roy et al. 1968: 17–18.

ogies and different innovations. In these situations, some researchers determine the technological appropriateness of each item for the individual farmer's situation, then produce an index based on the percent of applicable items used by the farmer; others do not.

Finally, both year of first use and index measures may vary according to whether they record trial or adoption of the practice in question. And, both trial and adoption may be broken down by the percentage of the farmer's fields to which they are applied, but this detail is rare in the literature. In the cases of the predominantly "successful" innovations studied here, trial is regularly followed by adoption; but in principle a farmer may try a new practice and then decide against continuing its use. It seems clear that trial, rather than eventual adoption, is the better operationalization of the theory presented in Chapter 2. It is used in this study whenever possible.

### *Farming Situations and Measures of Innovation*

The agricultural communities studied here cover an enormous range— from Maya Indians who farm hillsides with digging sticks and hoes to Wisconsin dairy farmers who depend on electricity and stainless steel. They include irrigated and unirrigated farms, large and small farms, specialized commercial farming directly serving cities, and relatively isolated subsistence farmers. In the discussion of economic rank in Chapter 4, some comparability was achieved through use of one uniform measure: land area. The differences between communities that are implied by differences in quality and extent of lands simply were not emphasized. Parallel masking of differences is more difficult in the case of the new practices studied here for each community. There is no common denominator like land.

Since each community is treated as a case, it is most important to make certain that internal comparisons are valid. Strict comparability among cases is not really needed for this analysis. Though the original investigators whose data is used here varied greatly in the attention they gave to reporting details of the local farming situation and the specific innovations whose spread they studied, I feel some detail is appropriate for two reasons. First, as an anthropologist I simply feel more comfortable knowing something concrete and specific about a situation included in a generalization. I do not mean to imply that every situation must be treated as forever unique, rather only that it is wise to know something of its unique characteristics. Second, this knowledge of unique or special characteristics, while irrelevant to overarching generalizations, is exactly what is needed for formulating alternative hypotheses, and for deviant case analysis once the original hypotheses have been tested. It is supplied for the record, and for the reader who wishes to reinterpret my results.

A brief narrative description of the farming situation and the basic measures of innovation for all the studies is found in Table 10. The details offered are not exactly parallel from case to case, though certain basic facts, like type of crops, are included for each case. Rereading the background works to uncover the details included here was, in most cases, a voyage of discovery on which I learned or relearned facts that I had not remembered from other readings over a period of four years. The resultant table will, I hope, give the reader a feeling for the range of situations covered in this comparative study, as well as specific details he or she might want about the individual studies. The original publications listed in Table 2 and Appendix D of course include much more detail.

### *Manipulation of the Data*

Selection of the practice or practices used to measure innovation is crucial. After that, treatment of the data is relatively

TABLE 10. *Farming Situations and Measures of Innovation*

1) India 185
Pedapulleru farmers devote more than 90 percent of their effort to rice farming on land 100 percent of which is irrigated during both wet and dry seasons. Two-thirds of the land is double cropped. The irrigation system quality is rated as average in a comparison with 13 other areas in Asia (IRRI 1975: 185). Landowners—as opposed to pure tenants—operated 90 percent of the farms larger than 4.8 hectares, the mean size of farm (Parthasarathy 1975: 47).
"Modern rice varieties were introduced into the village through the HYV in the dry season of 1965–66" (Parthasarathy 1975: 48). By 1971–72, 12 percent of farmers were using modern varieties in the wet season and 45 percent were using them in the dry season. "Since modern varieties planted in wet season mature during the period of heavy rainfall, Pedapulleru farmers lost interest in them" (Parthasarathy 1975: 49).
2) India 412
The 8 villages of India 412 display great variety within and across villages. Mixed farming is dominant, with about half the farmers getting half or more of their total value of production from a single crop (Data codebook). Sugar cane and potatoes seem to be the dominant cash crops (Codebook; Roy et al. 1969 do not discuss specific crops very much). Average farm size based on the entire sample that included 95 nonowners was one hectare in 2.5 different plots (Roy et al. 1969: 76). About 40 percent of the respondents said they used no irrigation and about 40 percent said their land was fully irrigated (Codebook).
12 agricultural innovations that were judged to be applicable to all villages and salient to villagers and change agencies were selected from a much longer list (Roy et al. 1969: 72) and used to make the index of adoption. Table 5.3 in the original publication (Roy et al 1969: 78–79) lists 13 innovations, and the Codebook records the adoption of a maximum of 11. The innovations include new seeds, techniques, implements, fertilizers, animal inoculation, and rat poison.
3) Kenya 540
"The economy of South Imenti in 1971–72 was basically agricultural, although almost half of the households reported additional nonfarm earnings. There were three distinguishable zones of agricultural production: the tea zone, the coffee zone, and the millet zone" (Almy 1974: 38).
The innovation variable was year of first use of insecticide on food crops (p. 56). (This item was chosen over 2 alternatives because of its distribution.)
4) Mexico 108, 5) Mexico 193, and 18) Mexico 93
Zinacantecos practice slash and burn maize and milpa agriculture on rented, rainfed, hilly land in the hot Grijalva River valley a day's walk from their homes in the cool highlands. Subsistence aspects of agriculture are crucial, but most farmers sell part of their crops and many sell much more than half.
New roads and government receiving centers built in the decade before

*Continued*

TABLE 10. *Continued*

1966 opened more productive lands to Zinacanteco farmers, who had little experience with climate and rainfall in the distant locations or with the complications of selling crops at the receiving centers. For Mexico 93 the innovation variable is year of first sale to the government receiving center. For Mexico 108 and 193 the innovation variable is distance from home of the farmer's fields in 1966.

6) Pakistan 221

"The usual rotation in the barani [rainfed] area is to grow wheat in rabi and maize, sorghum, and millet in kharif. Wheat is normally sown from October to mid-December depending on the autumn rains and harvested during April or May. . . . Seasonal labor demands are greatest during the short harvest and sowing period between kharif and rabi season" (Rochin 1971: 21). The area is poor and has rough terrain. The study concentrated on the spread of dwarf wheat in the area, for which it was not designed.

The innovation variable is year of adoption of dwarf wheat.

7) Pakistan 350

"Khanewal Tehsil is a major wheat-cotton area which is irrigated by perennial public canal water supplies" (Lowdermilk 1972: 21).

The innovation variable is year of adoption of dwarf wheat.

8) Philippines 153 and 9) Philippines 177

"In the Dumaguete trade area, most farming is a combination of cash and food crop" (Pal and Polson 1973: 83). Corn and coconut are the major crops. The land is not irrigated. The farms are cultivated by use of animal power (p. 18). In 1952, 55 percent of the families in the area had farming as their most gainful occupation; in 1966 the figure was 46 percent (p. 83).

The adoption index includes eight items for 1952 (N = 177) and 14 items for 1966. These include irrigation, chemical fertilizer, and green manure as the only items adopted by more than 10 percent of the overall (N = 307) population in 1952. Adoption rates for these and other practices were higher in 1966; but on the whole farmers adopted few of the total number of practices considered. In 1966, operators whose first or second most gainful occupation was farming had a mean adoption score of 3.38 of the 14 possible items (pp. 112–113).

10) Taiwan 159 and 11) Taiwan 237

"Shangfung and Liupao are in a predominantly rice producing area. Under normal circumstances, two rice crops are grown in a year with one or two intermediary crops between rice crops" (Lionberger and Chang 1968: 13). Wheat is also an important crop for many farmers. "Being 'at the tail of the irrigation water,' more farmers in Liupao grew such upland crops as wheat, peanuts, sweet potatoes, citrus fruit, and soybeans; fewer grew vegetables and rape seeds than in Shangfung" (p. 13).

The adoption index included information on 8 recommended practices for rice culture and a total of 21 others for five other crops. "All farmers who were growing rice were asked the rice culture questions. All others were

TABLE 10. *Continued*

asked questions about their most important commercial crop (the one that brought in the most money). In addition, each farmer was asked about practices used for a second crop, usually the second most important one commercially. His adoption score, then, was simply the percentage of the practices he had adopted of those included in the two sets of questions" (p. 23).

12) U.S. 159 and 13) U.S. 167

"The two communities from which data for this study were taken offered broadly contrasting conditions in farming as a way of life. Ozark [N = 167], located in hilly southern Missouri, is characterized by relatively poor soil and by low farm income in comparison to the state average. Prairie, the second community, is located in northwestern Missouri, where the land is fertile. . . . [They] grow corn, hogs, cattle and more recently, soybeans. There is little need in Prairie to supplement farm income by off-farm work and far fewer than in Ozark have had either to leave their farms permanently or to seek off-farm work because of economic necessity" (Lionberger and Copus 1972: 76). "Dairy production prevailed as a chief source of income in Ozark" (Lionberger and Campbell 1963: 9).

The Missouri and Taiwan studies conducted by Lionberger and associates were designed to be parallel, and the index used for Missouri is of the same type described for Taiwan 159 and 237.

14) India 676

Farming in each of the three states is described in Roy et al. 1968: 25–30.

*Andhra Pradesh.* "Like most package districts, West Godavari in Andhra Pradesh is bountifully endowed with natural resources and presents an attractive picture of a generally prosperous agriculture" (p. 25). "Reflecting the labour shortages, modern implements have some acceptance in the study villages" (p. 26). "Farming operations generally follow the typical Indian pattern of *khariff* (first crop season which begins in June-July) and *rabi* (second crop season which begins in October, November). In *khariff*, paddy is usually grown, although some farmers grow groundnuts and sugarcane. Paddy sugarcane, banana and chillies make up the main *rabi* crops . . . Water is supplied by both canal networks from the Godavari river and by filter point wells" (p. 26).

*Maharashtra.* "Agriculture is also the most important activity in the two sample Maharashtra villages . . . As in Andhra Pradesh, much labour is hired, and there are occasional shortages in peak seasons (p. 27). "Improved modern implements are not often observed. There is no assured supply of irrigation water . . . [the] cycle of crop production is somewhat similar to Andhra Pradesh in that *khariff* and *rabi* seasons are recognized. However, the *rabi* crop makes up much less of the total than in Andhra, principally because of the lack of irrigation facilities . . . *jowar* and cotton are the main crops. Pulses are sometimes sown mixed in the cotton fields. Wheat is principally grown in the *rabi* season" (p. 28).

*Continued*

TABLE 10. *Continued*

*West Bengal*. "There are three sample villages in West Bengal and they differ from those in the other two states in several ways. There is more crop specialization, much more land fragmentation, and there are fewer dependable sources of irrigation in a predominantly rice culture. The general pattern of farming operations, however, remains the same" (p. 29).

"We decided that 'ever having used' a practice was the best measure of adoptive innovativeness and selected this as *the* dependent variable for this phase of our study. All 10 items were used and scored as a simple unit-weighted index" (p. 24). The 10 items were similar to those used in India 412. In some cases (e.g., HYV seed) the "item" represented a different locally appropriate innovation for each state (pp. 15–16).

15) India 1,145

"The state of Andhra Pradesh (. . .106,000 square miles) . . . includes both extensive areas of dry cultivation on the inland Deccan plateau (Telangana region) and relatively well-developed and densely populated rice-growing areas centered on irrigation from the Godavari, Krishna and Penna (Andhra region)" (Gartrell 1977: 321).

"The innovation index was a count of the reported number of items of new technology tried (largely the 'Green Revolution' technology of new seeds, fertilizers, pesticides, etc.). Respondents first were asked open-ended questions as to whether they knew anything about various types of new technology (for example HYV seeds) and, if they said 'yes,' they were then asked to specifically name those they knew about. Of those they correctly identified, the ones they had tried were included in the index of innovations . . . In the scale used here, scores between 10 and 32 represent the top 21.9% of the distribution" (Gartrell 1977: 322n). Fewer than 5 percent of the respondents used more than 15 of the 32 innovations included in the index (Codebook).

16) U.S. 316

All the farmers run corn-dairy enterprises.

"Six agricultural innovations were selected to measure the dependent variable. These include: 1) the use of atrazine for control of weeds in corn; 2) wheel-track planting of corn; 3) green-chop feed for dairy cows; 4) the use of nitrogen side-dressing in corn; 5) the use of a bulk fertilizer (spreader attached to a truck); 6) the use of Vernal Alfalfa which is disease and fungus-resistant. All the practices had been introduced about 10 years previous to the study. The innovation index was formed by summing the number of innovations tried for each subject (range 0–6)" (Gartrell et al. 1973: 394).

routine, for it is clear that earlier adoption (or trial) of the crucial practice (or adoption of more practices at a given point in time) is more innovative. The relationship between degree of innovation and time is monotonic, almost by definition. Thus, the complexities introduced to the manipulation of the data on economic rank by the curvilinear hypothesis are for the most part avoided in the case of the dependent variable.

The theory specifies that Stage 1 shall be a stage in which uncertainty is high and Stage 2 a stage in which uncertainty is lower. In this study, Stage 1 is operationalized as the first 25 percent of the population to adopt and Stage 2 as the second 25 percent to adopt. The remainder of the population is left undivided. Thus the innovation variable, in the final stages of the crosstabular analysis, always has three values. The selection of 25 percent to define each of the first two stages was dictated in part by simplicity, in part by the desire to use as large as possible a Stage 1 in order to increase the cell expected frequencies when the total population studied is small.

An obvious alternative procedure would have been to follow the adopter categories defined and described by Rogers (1962: 159–68). This categorization takes off from the tendency for adoption to be normally distributed over time and uses five categories from earliest to latest as follows: innovators (first 2.5 percent to adopt), early adopters (13.5 percent), early majority (34 percent), late majority (34 percent), and laggards (16 percent). As Rogers points out (1962: 159), innovativeness is a relative concept and has a continuous distribution; thus, the adopter categories are a conceptual convenience, not a substantively meaningful division of the distribution. In this book the need to insure a large number of "innovators" made the 25 percent definition of Stage 1 preferable to the 16 percent that might have been adopted in accordance with convention.

More important, of course, is the likelihood that the distinction between innovators and followers will be different for different practices and/or settings. That is, some kinds of innovations may be such that "knowledge" is fairly complete after 15 percent of the population has tried them—either because of their characteristics or because of the characteristics of the setting; others may require 35 percent adoption to establish

TABLE 11. *Summary of Measures of Innovation*

| Study | Measure of innovation |
|---|---|
| 1) India 185 | First year used HYV rice |
| 2) India 412 | Number of innovations used: 12-item list |
| 3) Kenya 540 | First year used insecticide on food crops |
| 4) Mexico 108 | Distance of fields from home in 1966 |
| 5) Mexico 193 | Distance of fields from home in 1966 |
| 6) Pakistan 221 | First year used HYV wheat |
| 7) Pakistan 350 | First year used HYV wheat |
| 8) Philippines 153 | Number of innovations used: 14-item list |
| 9) Philippines 177 | Number of innovations used: 8-item list |
| 10) Taiwan 159 | Percent of applicable innovations used |
| 11) Taiwan 237 | Percent of applicable innovations used |
| 12) U.S. 159 | Percent of applicable innovations used |
| 13) U.S. 167 | Percent of applicable innovations used |
| 14) India 676 | Number of innovations used: 10-item list |
| 15) India 1,145 | Number of innovations used: 32-item list |
| 16) U.S. 316 | Number of innovations used: 6-item list |
| 17) Japan 91 | Number of innovations used: 4-item list |
| 18) Mexico 93 | First year sold to receiving center |
| 19) U.S. 173 | Percent of applicable innovations used |
| 20) U.S. 251 | First year used hybrid corn seed |
| 21) U.S. 252 | First year used bluestone lime (for tobacco) |
| 22) U.S. 341 | Percent of applicable innovations used |
| 23) U.S. 423 | Number of innovations used: 9-item list |

familiarity. For example, where land is uniform across all farmers in a community, fewer "trials" are necessary than where the community works in a number of microenvironments. Or, the physical, social, or cultural setting may lead to different communication patterns that in turn affect the degree of information spread after any given proportion of the population has adopted. For example, the knowledge of nonadopters in the community may be as great in one place after 10 percent have adopted as it is in another place after 20 percent have adopted.

In addition to these fairly concrete problems, there remains the general problem that there is no clear theoretical reason to select any particular dividing point between the stage of uncertainty and the stage of adequate knowledge. In an ideal situa-

TABLE 12. *Ranges of Innovation Scores by Stage*

| Study | Stage 1 | Stage 2 | Remainder |
|---|---|---|---|
| 1) India 185 | 1965–66 | 1967–68 | 1969–72 |
| 2) India 412 | 6–11 | 4–5 | 0–3 |
| 3) Kenya 540 | 1938–67 | 1968–70 | 1971–74[a] |
| 4) Mexico 108 | 6[b] | 3 | 2 |
| 5) Mexico 193 | 8–9[b] | 6–7[b] | 2,4,5[b] |
| 6) Pakistan 221 | 1969–69 | 1969–70 | not yet |
| 7) Pakistan 350 | 1966–67 | 1967–68 | 1968–71 |
| 8) Philippines 153 | 5–9 | 4 | 0–3 |
| 9) Philippines 177 | 3–5 | 2 | 0–1 |
| 10) Taiwan 159 | 70–100% | 60–69% | 0–59% |
| 11) Taiwan 237 | 90–100% | 80–89% | 0–79% |
| 12) U.S. 159 | 60–89%[c] | 50–59% | 0–49% |
| 13) U.S. 167 | 80–100% | 60–79% | 0–50% |
| 14) India 676 | 7–10 | 5–6 | 0–4 |
| 15) India 1,145 | 10–32 | 7–9 | 0–6 |
| 16) U.S. 316 | —[d] | —[d] | —[d] |

[a] Includes nonadopters.
[b] Numbers indicate zones that measure distance (see Cancian 1972).
[c] No respondent used all the applicable practices.
[d] Measure actually a summary sten score for the 6 items.

tion we might want to compare, say, the fifth percentile and the fortieth percentile farmers on the innovation variable, thereby avoiding all the problems associated with the boundaries of stages. However, since no compelling reason for any one or two particular alternative manipulations of the data present themselves, the 25-25-remainder category system that has so many practical advantages is used here. There is more detailed discussion of alternative procedures in Appendix B. The specific rules for data manipulation are in Appendix A.

Table 11 shows the type of measure of innovation used in each case, and Table 12 shows the range of each of the three categories into which it has been divided.

### *Summary*

The innovations studied here are the new and different farming technologies and practices involved in the Green Revolution and other efforts to modernize farming in the United States and the third world. Measures of innovation (early adoption relative to others in the community) are developed on the basis of research by rural sociologists and others. Some of the measures are based on the year of first use of a crucial innovation (like new seeds for corn, wheat, and rice); others on the number of new practices (among many available ones) used at a given point in time.

The unknown quality of most new technologies and practices, and the resultant uncertainty about the implications of adoption, are central to the approach taken in this study. In keeping with the theory presented in Chapter 2, the measures of innovation are divided to provide a distinction in degrees of uncertainty. Stage 1, or the first 25 percent to adopt, is used to represent the period when uncertainty is high. Stage 2, or the second 25 percent to adopt, is used to represent the period when knowledge about the new technologies and practices has spread in the local community and the uncertainty is substantially reduced.

## *Chapter Six*

# Testing the Hypotheses

Testing the hypotheses in a study of this kind is not a straightforward procedure. The "pure" hypotheses stated in Chapter 2 must be tested with the "impure" data described in Chapters 3, 4, and 5. The reader who has skipped ahead to the conclusions in Chapter 7 knows that the assembled data strongly support both principal hypotheses; but the road from the raw and mechanical tests of the hypotheses, through understanding of the complications, to clear support for the theory and its various implications is not without detours. The process is at times more artful than scientific. Thus, I have tried to display the results fully so that the reader can make his or her own judgment of the conclusions I make.

The chapter is organized around two large figures. The first displays one form of all the crosstabular tests of the hypotheses. The second displays the evidence that remains after various complications—especially those concerned with identification of the appropriate community of reference—are resolved. The tests are presented immediately after a brief review of the hypotheses.

### *The Hypotheses Reviewed*

The two hypotheses stated in Chapter 2 are:

HYPOTHESIS A: *In the early stages of the spread of an innovation, low-middle-rank individuals are more likely to adopt it than are high-middle-rank individuals.*

HYPOTHESIS B: *In the later stages of the adoption process, the adoption rate of high-middle-rank individuals will increase relative to the adoption rate of low-middle-rank individuals.*

Hypothesis A is quite straightforward. Given measures for the ranks, it says simply: LM > HM, where LM and HM represent the percentage of farmers of each rank who adopt the innovation during Stage 1 of the adoption process. It predicts a negative section of some sort in the curve describing the relation to rank and innovation.

Hypothesis B is not quite so straightforward. It can be formally stated as $HM_2 - HM_1 > LM_2 - LM_1$, where the subscripts indicate stages; but this statement leaves two ambiguities that could effect the outcome of tests. First, is the Stage 2 adoption rate to be figured by dividing the farmers who adopt in Stage 2 by the total number of farmers in the rank, or by the number of farmers who have not adopted in Stage 1? Second, how are we to handle distortions created by lumpiness in the dependent variable? These complications are discussed and resolved in Appendix A. The reader who is not interested in the technicalities should know that they affect the tests of Hypothesis B only in a few cases: three of the forty-nine in Figure 8 and one of the sixteen in Figure 9.

### *The Display of Simple Crosstabulations: The Mechanical Analysis*

Figure 8 displays one or more histograms for each of the studies. The order of presentation is that followed in previous chapters; the newly collected thirteen studies, followed by the three studies on which published replications are based, followed by the seven studies from my original paper (Cancian 1967). In accordance with procedures discussed in Chapter 4, separate results are displayed for the different independent variables (land area and "income") where they are available in the original data, and for different cutting points (quartile and "30/20") where the original data permit them.

For seven of the first sixteen studies both independent variables and both cutting points are usable, so each study is represented in the figure by four histograms. Five of the studies are limited to one independent variable, but permit use of both

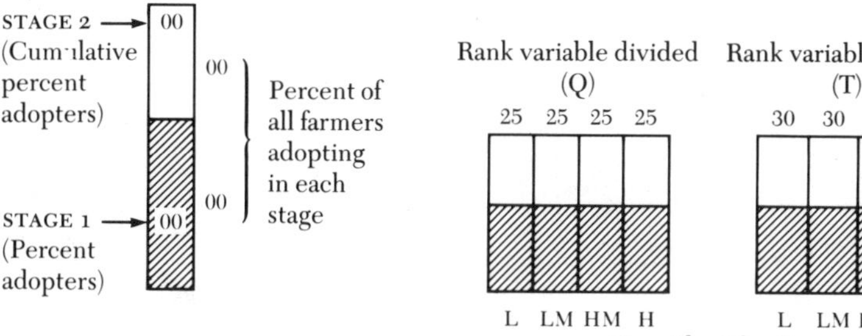


Explanation. The first graph to the right is to be read as follows. No individuals of low rank adopted in Stage 1, but 33% of the group adopted by the end of Stage 2; in the low-middle rank, 15% of the individuals adopted in Stage 1, with the total rising to 52% by the end of Stage 2; etc. In the sample of 185 individuals, 22% adopted in Stage 1 and 30% adopted in Stage 2, making a total (not shown) of 52% by the end of Stage 2. The label Area Q indicates that the independent variable is area of land (see Table 5), and that the 185 farmers are divided to approximate quartiles.

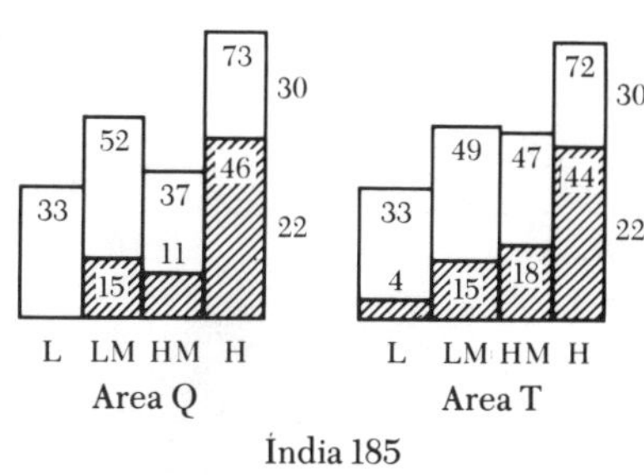


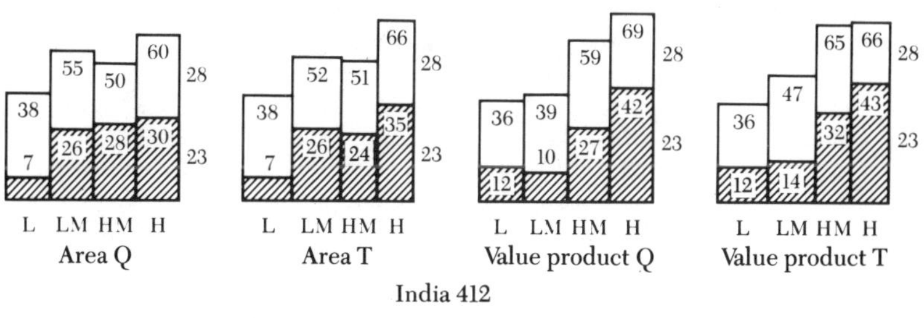


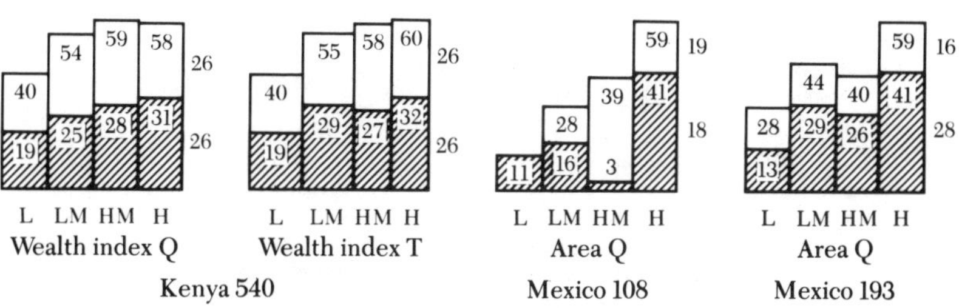


***Fig. 8. Extensive tests of hypotheses***

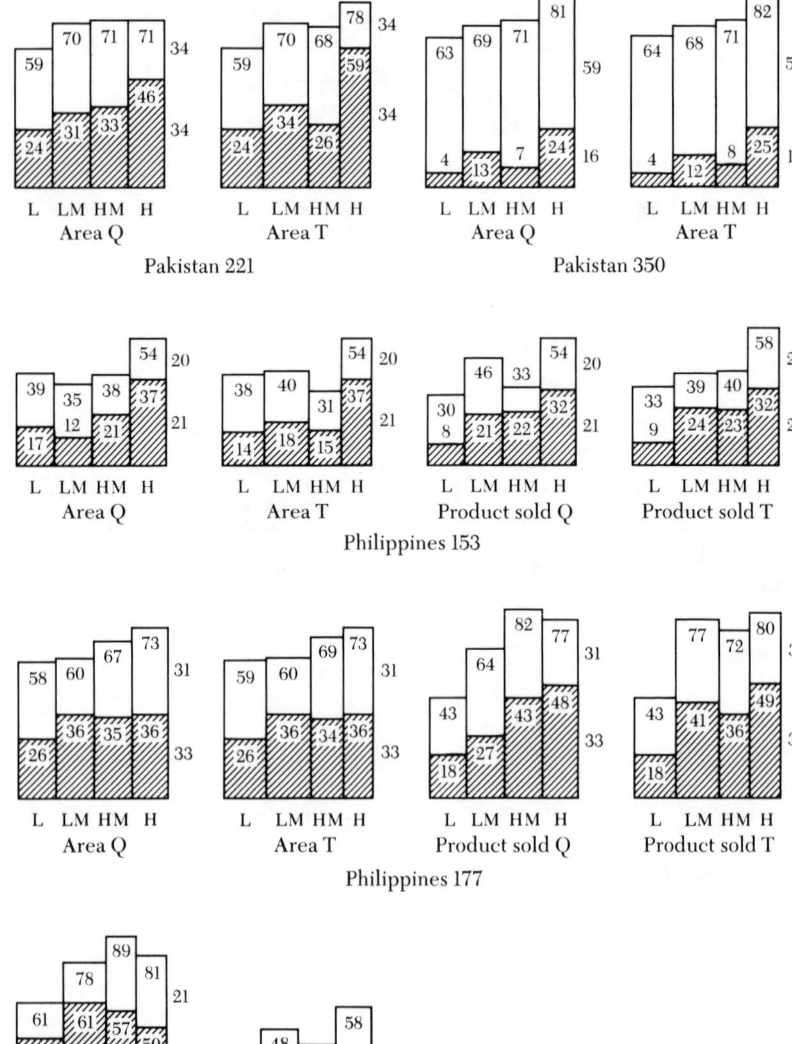


***Fig. 8. Continued***

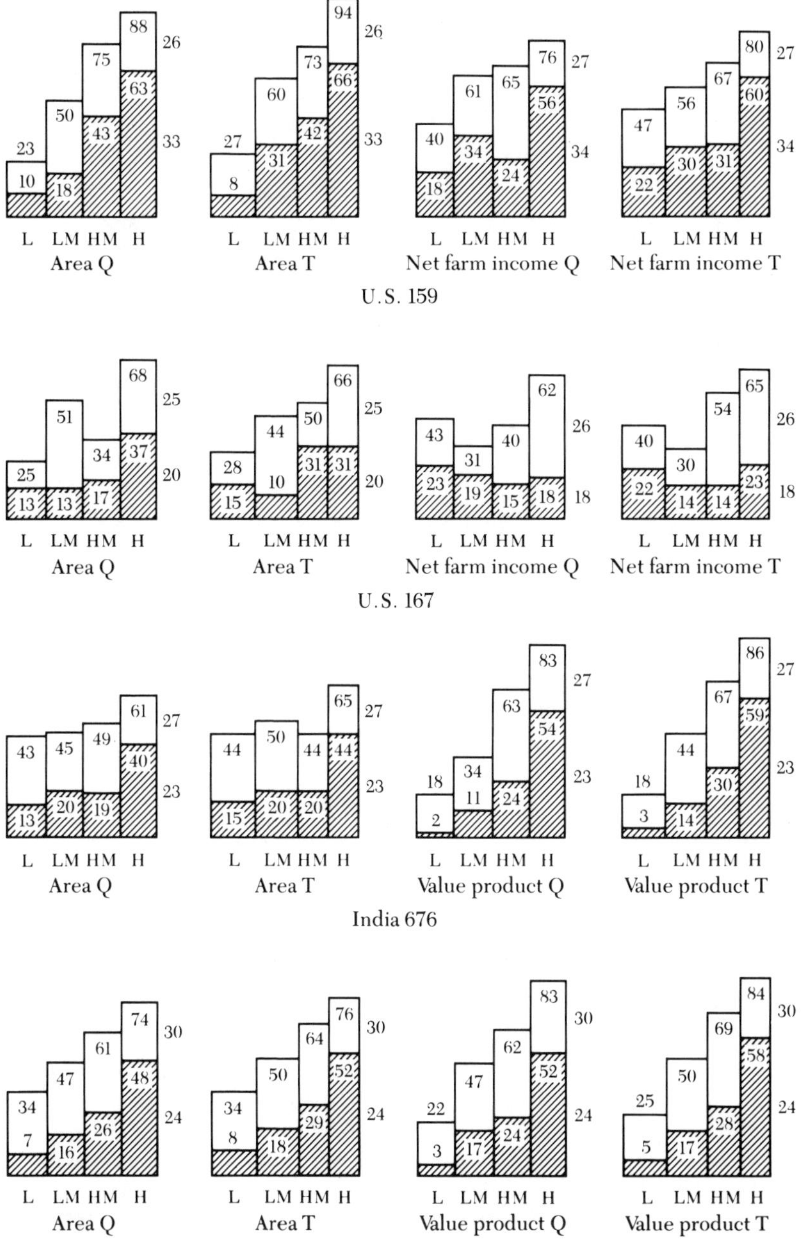


***Fig. 8. Continued***

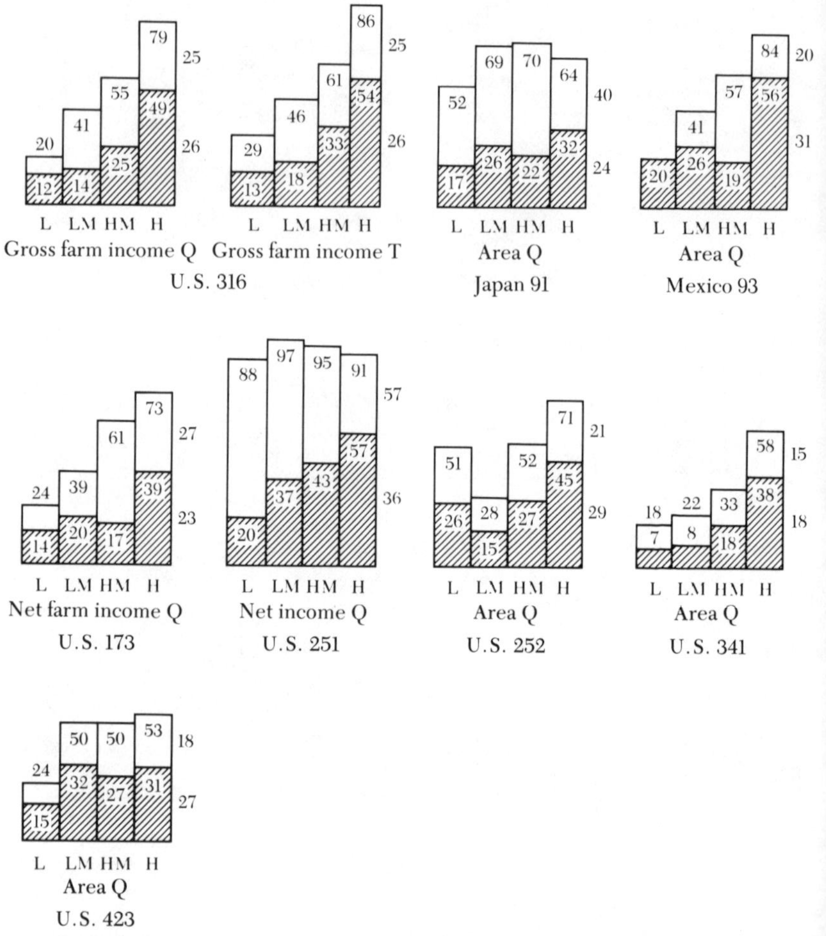

L LM HM H
Gross farm income Q

L LM HM H
Gross farm income T

U.S. 316

L LM HM H
Area Q
Japan 91

L LM HM H
Area Q
Mexico 93

L LM HM H
Net farm income Q
U.S. 173

L LM HM H
Net income Q
U.S. 251

L LM HM H
Area Q
U.S. 252

L LM HM H
Area Q
U.S. 341

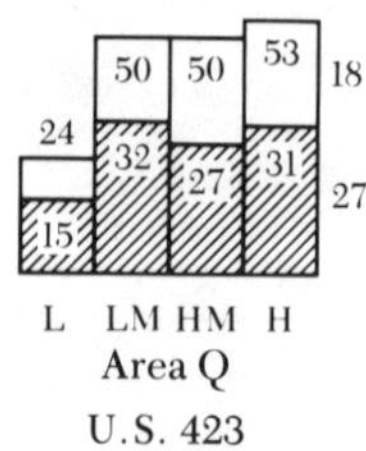


Area Q
U.S. 423

***Fig. 8. Continued***

cutting points; and five studies permit only a single independent variable-cutting point combination. Thus, these sixteen studies account for the first forty-two histograms. Each of the seven studies from Cancian 1967 is represented by a single histogram. The histograms for these seven studies are based on the earlier analysis of the data, and full tabular versions are in the article (Cancian 1967: 922). The full tabular version of each histogram for the first sixteen studies is found in Appendix A along with a display of "decile" cuts. Unstandardized figures are presented here to illustrate the lumpiness of the data. Standardized figures and decisions on the tests are recorded in Appendix A.

*This display is raw data.* Except in the case of Kenya 540, where Almy explicitly argues that land area gives a poor measure of economic rank (see Chapter 4), both area and "income" variables are presented whenever they are available.

It is easy to argue different conclusions from this raw data. The following illustration of the range of conclusions that can be argued from the data suggests that no conclusions be drawn until the data are carefully considered. On the one hand, only 23 of the 49 histograms shown in Figure 8 support Hypothesis A. One could conclude that the hypothesis is not supported by the evidence. On the other hand, a few very simple and reasonable steps radically change the picture: the three "negative" cases from the literature represent the worst approximation to randomly selected cases among the assembled cases. Eliminating them immediately switches the balance, so that 21 histograms support Hypothesis A while only 18 do not support it. Further elimination of the studies done in the United States, on the grounds that they represent a special situation very different from the peasant farmers typical of the other cases, changes the count to 17 to 9 in favor of Hypothesis A. Eliminating the quartile cuts of the independent variable (the Q histograms) where 30/20 cuts are available (on the basis of the arguments of Morrison et al. reviewed in Chapter 4) makes the count 14 to 2 in favor of Hypothesis A. Finally, concentration on the area variable where both area and income variables are available eliminates "double counting" of three cases and makes the final tally 12 to 1 in favor of Hypothesis A. A similar exercise for Hypothesis B takes the count from 25 to 24 to 11 to 2.

*Neither the conclusion that Hypothesis A is not supported, in 26 of 49 cases, nor the conclusion that Hypothesis A is supported, in 12 of 13 cases, is warranted as far as I can tell.* The discussion that follows shows that the best approximation to the truth lends very substantial support to the hypotheses.

### *The Less-Mechanical Analysis*

The less-mechanical analysis of the available data must consider a variety of factors that may influence the outcome of any particular test. These include:

1) selection of the cases from the universe of cases
2) definition of the community of reference in each case
3) selection of the cutting points for the independent variable
4) type of independent variable
5) selection of the cutting points for the dependent variable
6) type of dependent variable
7) other and idiosyncratic factors

Below I shall discuss each item in this list. Since most of these factors can significantly influence the outcome, it is difficult to assess the effect of one without simultaneously assessing the influence of the others. Thus, conclusions about the analysis should be postponed until all the factors have been reviewed.

### *Selection of the Cases from the Universe of Cases*

The theory tested here is meant to apply to every community of farmers (or other people). The twenty-three cases included in the analysis were not selected at random (Chapter 3) and are not meant to be representative of any particular universe of communities. Nor is there any reason to think they are atypical in any way that affects the tests.

Thus, the argument made above, for elimination of the three negative replications on which publications are based on the grounds that they are not randomly selected, is inappropriate. In fact, each of these cases should be eliminated from consideration for other reasons stated below, so the substantive result of eliminating them does not change. Nevertheless, it is imporant to recognize that they are not eliminated for sampling reasons. Were criteria of randomness applied, the positive Zinacantan cases (especially Mexico 108 and Mexico 93, which represent readings taken on the same community two years apart) should be eliminated; and the inclusion of both Philippines cases might be questioned because they represent readings on the same area taken 14 years apart.

In the same spirit, the elimination of all the studies done in the United States that was suggested above is unwarranted, I think. While there is no doubt that most of the United States studies concentrate on farmers who are comparatively rich and involved in the use of relatively advanced technology, while the other studies concentrate on populations that are for the most part at the other end of both the wealth and technological complexity scales, the theory provides no reason to think that the relation between ranks is substantially different in the different societies. Thus, insofar as we are testing the theory, the United States studies should be no different from the other studies. They should not be eliminated on the grounds that they represent vastly richer and technologically more complex farming operations. Rather, the similar implications of the theory for both rich and poor communities of reference should be emphasized. In sum, no cases may be eliminated on the grounds that they are unrepresentative of the populations about which I would like to generalize.

### *Definition of the Community of Reference in Each Case*

The theory stated in Chapter 2 assumes that the behavior studied takes place within a community in which the actors (or classes of actors) are relevant to one another. Chapter 3 discusses the complications of operationalizing such a notion. In practical terms, it is not appropriate to test the theory on a population drawn from two or more communities of reference. This is because the distribution on, say, the independent variable might be different in the different communities—making a rich farmer in one community a poor one in another, or vice versa. Thus the relative rank positions of the farmers would be inaccurately measured if the communities of reference differed in this manner.

Just how big a community of reference is varies from place to place, of course. Landscape, population density, and communication systems, among other things, may influence the size of the effective community of reference. While it is always possible for a large area to be homogeneous, so that measures taken from different subpopulations (communities of reference) within it are comparable, the larger the population the greater the chance that subpopulations will not be comparable.

The actual population size of the sampled universes for the

TABLE 13. *Universe Size in the Twenty-three Studies*

| Size | Number of cases |
|---|---|
| Less than 10,000 | 9 |
| 10,000 to 100,000 | 6 |
| 100,000 to 1,000,000 | 3 |
| More than 1,000,000 | 5 |

NOTE: Table 4 shows the universe size for each of the 23 cases.

twenty-three cases in this study varies from less than one thousand to more than one hundred million (see Table 4). Table 13 summarizes the distribution of the twenty-three cases over the wide range of population sizes.

It seems appropriate to consider eliminating the five studies from universes of more than one million people. In two of these, India 676 and India 1,145, the authors explicitly report efforts to include diverse subareas in the sample. Another (U.S. 316) attempts to cover all of rural Wisconsin, and a fourth (U.S. 423) covers a large part of North Carolina. Appendix C discusses efforts to analyze the subpopulations represented in these samples.

In the course of examining the fifth case (Pakistan 221), Appendix C shows that two universes of approximately 25,000 can be identified within the original universe of more than 1,000,000. Thus, while Pakistan 221 should be removed from the data sets used to test the hypotheses, Pakistan 105 and Pakistan 116 may be added.

There are only three cases in the 100,000 to 1,000,000 category in Table 13. They are India 412, Pakistan 350, and U.S. 341. The largest universe, for U.S. 341, is estimated at about 350,000 people. India 412 is estimated at 200,000 and Pakistan 350 at 185,000. These are large populations that encompass many villages or many counties. Without information definitely suggesting unusual homogeneity across the area involved, it is, in my judgment, inappropriate to assume that they constitute single communities of reference or multiple communities with similar distributions on the important variables.

The six cases taken from areas with populations between 10,000 and 100,000 are, it seems to me, all good approximations of communities of reference. The three from the United States (U.S. 173, U.S. 251, and U.S. 252) are all within one county. The two from the Philippines represent the farmers around a single market city; and the Kenyan case represents a single ethnic group in a single area. This judgment that these particular cases are appropriate is based on the information displayed in Tables 3 and 4 and in the original publications. Obviously, it would be possible to encompass considerable diversity in a universe of less than 100,000 people. That is, population size alone cannot be routinely used as a criterion for community of reference.

Table 4 shows that the nine cases taken from areas with populations of less than 10,000 are in fact censuses of all (or virtually all) farmers in a natural political or social subdivision (village or town) with a population of less than 2,000. The problem, if any, with these cases might be that they fail to represent a substantial part of the relevant community of reference. The supplementary information on them gives no indication that this is true for any of them. Thus they all seem to provide appropriate tests for the hypotheses.

In sum then, it is probably appropriate to eliminate the eight cases taken from areas with populations of more than 100,000, to add Pakistan 105 and Pakistan 116, and to continue analysis of the seventeen cases that can be identified as coming from areas with populations of less than 100,000. This important manipulation of the data provides an operationalization of the community of reference notion developed in Chapter 3.

### *Selection of the Cutting Points for the Independent Variable*

The clear prediction that the lower middle rank will innovate more than the upper middle rank in the early stages of the spread of an innovation is meaningless without specification of the ranks. Chapter 4 reviewed a number of issues relating to the specification of the four ranks necessary to test the theory. Appendix B discusses the complexities of continuous measures that might aid an inductive search for the predicted dip and provide a more suitable test of the hypothesis.

Figure 8 presents the results of using both cutting points specified in Chapter 4. The division into quartiles is favored for its simplicity, and the 30/20 division is favored for its better fit to common notions of the sociologically significant stratification continuum. If it were clear that the quartile or the 30/20 division were better on theoretical grounds, it could be argued that one rather than the other should be used for testing the hypotheses. Since it is likely that the location of the dip will vary from community to community for as yet unspecifiable reasons, it is difficult to choose.

Figure 8 shows that the 30/20 cutting points provide more confirmations on the cases where it is possible to choose between them and the quartile division. Among the first thirteen cases (India 185 through U.S. 167), the 30/20 cut favors Hypothesis A where the quartile cut does not in the following cases: India 412, Kenya 540, Pakistan 221, Philippines 153, and Philippines 177. The quartile cut is advantageous to Hypothesis A for India 185, U.S. 159, and U.S. 167. Pakistan 350 confirms the hypothesis with both cutting points.

No definite conclusion can be made because the numbers are too small. Both the number of cases (studies) and the number of individuals (observations) involved in most of the cases is too small to support any conclusions about the difference between the two cutting points. I speculate that it may be significant that the three cases where the quartile cut is favorable to the hypothesis are probably the richest among the eight cases involved. Table 6 shows that the farmers in these three communities have the largest farms on the average (although India 185 is clearly in the range of the non–United States communities, while the other two are United States communities); and Table 10 suggests that the Pedapulleru (India 185) land is of high quality. This pattern is consistent with the fact that the farmers in these communities are probably the best equipped to meet the absolute cost of new farming techniques.

In sum, the crosstabular tests of Hypothesis A suggest that the 30/20 cut may provide a better chance of locating the predicted dip in the distribution of farmers, especially if the community is poor on a world scale. This conclusion is consistent with the theoretical reasoning of Morrison et al. (1976), which I find convincing. For the purposes of the present study, however, it results from inspection of the raw data presented in Figure 8 (that is, from induction), not from any prediction made from the theory presented in Chapter 2.

### *Other Thoughts Relevant to the Interpretation of Figure 8*

The remaining ideas relevant to the interpretation of Figure 8 are of two kinds. First I will discuss points 4) through 6) above, though the variance available to support speculation about them is not great. Second, the richness inherent in any natural situation involving important economic activity produces a variety of idiosyncratic configurations and ad hoc explanations, even when the review of such situations is filtered through the original study, the secondary analysis, and the reshaping necessary to make it comparable to other similar studies. I have tried to keep the number of ad hoc explanations down—in part because I am sure that readers who have taken the trouble to get this far in the book have ideas of their own generated by combinations of information and sources of insight not available to me.

*Type of independent variable.* As noted in Chapter 4, there are reasons why net farm income might be expected to be a better measure of economic rank than area of land farmed. Gross farm income should also be a better measure than area farmed, except perhaps in mixed farming areas or areas including different specialties. In the cases in hand the area measures favor the hypotheses slightly more often, except in the cases of U.S. 159 and U.S. 167, a pair of Missouri communities where a net rather than gross income measure is available.

*Selection of the cutting points for the dependent variable.* Since the cutting points on the dependent variable were not varied systematically, as the cutting points on the independent variable were, there is little to be said beyond the *a priori* considerations discussed in Chapter 5. They do lead to the conclusion that at some point between 25 percent and 50 percent we can expect the curve describing the relation of rank and innovation to tilt up. That is, in fact, what Hypothesis B predicts. Thus, if the operationalization of Stage 1 in any particular case includes an exceptionally large percentage of the farmers, we may expect failure to confirm Hypothesis A. I believe this is what happened with U.S. 251, the famous Ryan and Gross study of the diffusion of hybrid corn seed in Iowa. Both the income rank variable used here (because it gives better approximations to quartiles) and in Gross' master's thesis (1942: 96), and the area rank variable used in the major published report (Ryan and Gross 1950: 691) and in the thesis (Gross 1942: 97), show that the twenty-three farmers (less than 10 percent) who adopted before 1934 lend more support to Hypothesis A than do the cumulative figures through 1936 (more than 36 percent), which it is necessary to use because of the way the data are reported. The numbers are too small to mean anything in themselves, but the pattern of their variation over the adoption process does support the speculation that the U.S. 251 case might have an inappropriately large Stage 1.

*Type of dependent variable.* The type of dependent variable (see Table 11) cannot be shown to make much difference to the outcomes in Figure 8, though I retain my feeling that an index based on a number of innovations can be problematic when mixed farming or different specialties exist in the study area. It is worth noting that the year of first use of insecticide was chosen as the dependent variable in the Kenya study because its distribution fits better with the other year of first use variables (see Chapter 5). Both year of first use of row planting and year of first use of fertilizer, which are available for the Kenya study and are only slightly less appropriate measures than year of first use of insecticide, yield support for Hypothesis A for both the quartile and the 30/20 cuts, while year of first use of insecticide does not confirm the hypothesis with the quartile cut.

*Other and idiosyncratic factors.* The distribution of resources on the absolute scale underlying the independent variable may also have an important influence on the definition of the situation within which the farmer operates. The theory tested here is based on rank-seeking behavior, and it seems reasonable to argue that extreme distributions might lead to a reduction in rank-seeking through agricultural innovation. Extremely equal distributions, some would argue, would make rank differences phenomenologically trivial, and thus they would not be effective predictors of behavior. Extremely unequal distributions, some would argue, might have a similar effect because either 1) they signify a rigid rank system or, to put it another way, 2) the difference between farmers is so great that individuals lose hope of gaining or fear of losing rank. Unfortunately I have found no way to make use of these arguments in interpreting the data presented here. It seems clear that the Mexican communities which have the most equal distribution of resources (see Gini coefficients in Table 6) are unequal enough to exhibit behavior patterned on economic rank. It may be that the most unequal community (India 185) is affected by the distribution of wealth, but I see no consistent pattern worth pursuing with the data resources at hand.

Finally, I would like to draw attention to the high rank and the low rank, which have been silent companions of the middle ranks throughout this enterprise. They have behaved consistently, the low rank showing a low and the high rank showing a high innovation rate. In a sense, they constantly affirm the importance of the facilitating effect of wealth. Of the many ad hoc explanations for the few exceptions to these generalizations about the low and high ranks, one seems worth repeating here. Eugene Anderson has suggested that the low adoption rate of the high-ranking people in the Taiwan cases may be explained by a tendency of truly innovative high-ranking people to seek economic success in the nearby cities. This explanation has the virtue of reintroducing both individual characteristics and the larger economic system. While making no substantial effort to determine their importance, the present study does not mean to deny that importance. At this point in history it is almost certainly true that individual characteristics have received too much attention and that the alternatives offered by the larger system, and the constraints imposed by it, have received too little attention.

### *Summary and Final Tabulation of the Test Results*

The principal hypotheses about the relationship of rank and adoption of new agricultural practices were tested in this chapter. Figure 8 displays the extensive analysis of all twenty-three data sets. A number of factors that complicate the analysis were reviewed; and the cases more likely to represent relevant communities of reference were identified. The latter procedure led

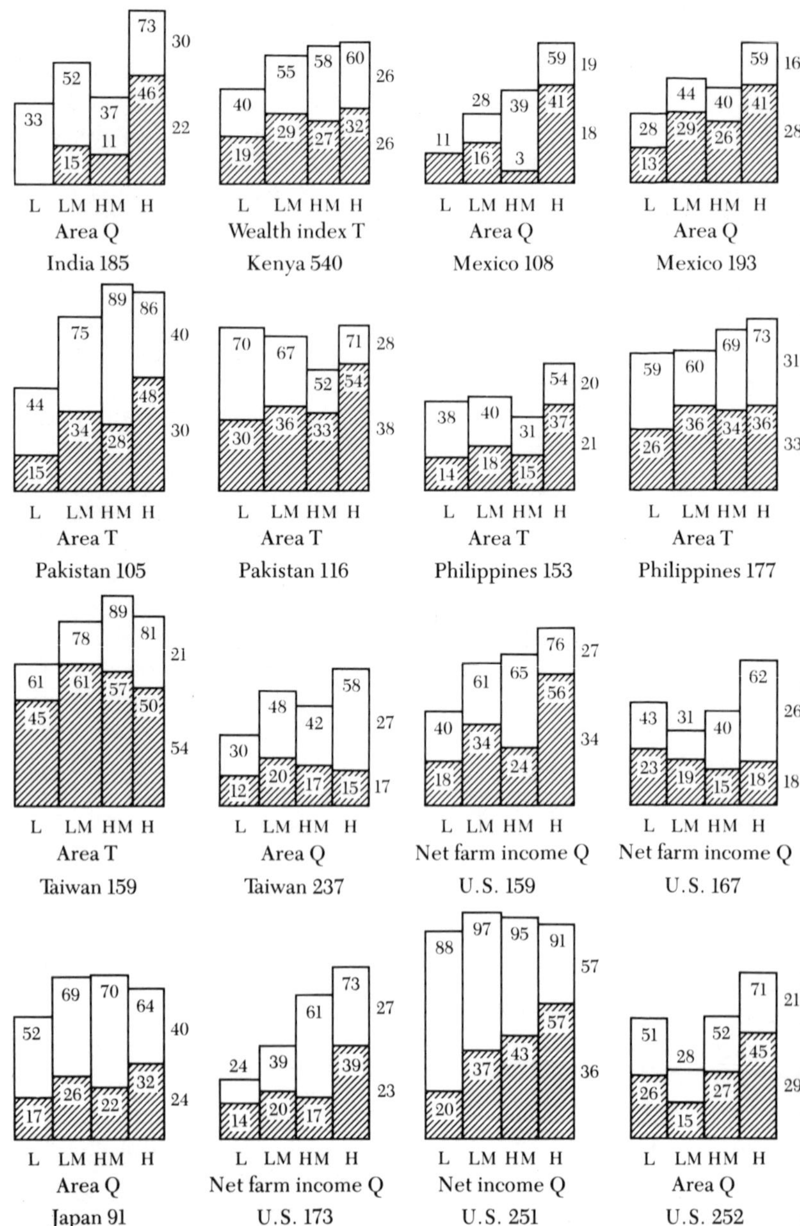


***Fig. 9.*** *The sixteen selected tests of hypotheses*

to elimination of seven data sets (studies) because they were sampled from universes of more than 100,000 people; and to the identification of two cases within the Pakistan 221 case. Thus application of the community-of-reference notion to the available data resulted in identification of seventeen useful data sets.

Tabulation of the results for these seventeen data sets that represent sociologically isolatable communities of reference shows support for both principal hypotheses. When the best approximation to the area independent variable and the 30/20 (T) cutting point are used for each case, 12 of the 17 cases support Hypothesis A and 11 of the 17 cases support Hypothesis B. While the area variable and the 30/20 (T) cutting point represent the most appropriate uniform method of analysis, other arguments made in the chapter suggest that: the quartile (Q) cut of the net income variable be used for U.S. 159 and U.S. 167; the quartile cut be used for India 185; and Mexico 93 be eliminated. These changes yield the final selection of sixteen relevant cases in Figure 9. Fourteen of the 16 support Hypothesis A and 12 of the 16 support Hypothesis B.

In sum, both hypotheses are strongly supported by the data. The analysis also illustrates the crucial importance of the community-of-reference notion. The broader implications of these results are discussed in the next chapter.

*Chapter Seven*

# Conclusions and Implications

The conclusions of this study fall into two major parts. One concerns the shape of the relation of rank and innovation. The other concerns the importance of the actor's relative social position described by the terms rank, community of reference, and innovator's situation. In this chapter these conclusions will be briefly described, and then their implications for the study of stratification and for agricultural development policy will be laid out.

### The Relation of Rank and Innovation Over Time

The theory and predictions stated in Chapter 2 are confirmed by the findings of this study. These conclusions can be summarized in three sentences. 1) It is useful to distinguish between an early stage in the spread of an innovation when uncertainty is high (Stage 1) and a later stage where knowledge about the implications of using the new practice is much more widespread (Stage 2). 2) In the first stage of the spread of an innovation, farmers of high middle position are apt to be quite conservative compared to those of low middle position in the community. 3) In the second stage of the spread of an innovation these high-middle-rank farmers catch up so that the overall relation of economic rank and adoption becomes monotonic, positive. That is, the relation of rank and innovation varies in a predictable way over time. Figure 4 illustrates the changing shape of the relationship over the first two stages of the spread of an innovation.

The findings resolve the apparent contradiction between the prediction that the relation of rank and innovation "dips" in the early stages of the spread of an innovation and the received wisdom that the relation is monotonic. The confirmation of Hypothesis B indicates that the dip present in Stage 1 tends to be replaced by a monotonic relationship in Stage 2. Thus both the dipping and the monotonic relation exist at different times in the process of the spread of an innovation.

These findings are quite general. They apply to Maya Indian corn farmers in Mexico, to tea and coffee farmers in Kenya, to Pakistani wheat farmers, Taiwanese rice farmers, Japanese truck farmers, and Wisconsin dairy farmers, among others. There is thus reason to believe that they apply to most if not all agricultural populations. This much can be said on the basis of the present study. I hope that, in some other place, the theory will be generalized to cover nonagricultural populations, and innovations and resources other than economic ones.

On the other hand, the findings are quite limited. They are shown to apply only to populations distinguishable as communities of reference, and only when the ordinal nature of the rank variable is emphasized. These restrictions are, of course, additions to the theory and conclusions of this study. They extend the theory and make it more useful. Their positive implications are the main subject of the following section.

### *Rank, Community of Reference, and the Innovator's Situation*

The findings depend on the idea that economic rank is a social variable, and they in turn support the reconceptualization of economic rank employed in this study. The social nature of economic rank within a community of reference is not a simple, direct implication of the findings. It is supported by interpretation of the overall pattern of results. It proves to be a useful preconception. Here I will briefly discuss "rank," "community of reference," and "innovator's situation," the three interrelated terms used in the text to specify the reconceptualization involved in using traditionally economic measures as social variables. That is, I will briefly explore the complexity of the independent variable used above.

The variable is rank—that is, relative position in a hierarchy —not absolute amount of the relevant resource controlled. This is social because it emphasizes relationship to other people and downplays the importance of material or other resources. The findings include two major elements that support the relevance of relative rank rather than absolute standing. 1) The patterned relation of rank and innovation is shown to hold across a number of very different absolute levels of wealth. That is, the bends in the curve describing the relation of rank and innovation fall at different points measured on an absolute scale. 2) The greater success of percentile ranking in the regression analysis (see Appendix B) suggests that absolute wealth measures (like land and cash income) are less useful calibrations of the differences between individuals in the stratification system.*

Second, the specification of a community of reference adds to the social nature of the independent variable expressed by rank. The rank concept is used to call attention to the social (relative) nature of the independent variable, that is, to the way in which it reflects relations to other people, not relations to absolute quantities of material or other resources. The community of reference notion is used to call attention to the social (group) nature of the independent variable: to the way that the community as a whole has an internal structure. In the theory elaborated in Chapter 2, positions at the ends of the rank continuum are conceptualized as different from positions in the middle. More and less rank does not provide the implications necessary to specify the theory. The curvilinear pattern of the relationship of rank and innovation is interpretable only when the points in the rank continuum can be defined relative to the ends. While this social nature of the independent variable that is expressed by "community of reference" shares much with the ideas that normally flow from the concept of "rank," it is useful to separate it.

Third, the independent variable in each case measures the individual's situation in his or her society. This aspect of the independent variable connects the individual's action, based on a postulated desire to attain high rank, to his or her rank in the local system. Discussion of the actor's rank as the innova-

*While it is technically correct that the percentile transformation of the independent variable could be approximated by an interval (cardinal) transformation of some kind, the ordinal transformation makes both practical and theoretical sense.

tor's situation emphasizes the actor's point of view on the same facts described by the combination of the rank and community-of-reference notions. I hope it enhances the recognition that both standing relative to other people and joint local definition of the socially relevant universe are important influences on the adoption of new farm practices and other innovative processes.

Some readers will classify this effort at theory construction and explanation of behavior as social-psychological or even psychological because of the statement of the problem in terms of the innovator's situation and my casual postulation of motives and desires on the part of actors. (On the other hand some will see the independent variable as an economic variable because it is economic rank.) Thus it seems worth emphasizing again that the measurement in the end is always of *position relative to other people in a defined group*. The independent variable is never conceptualized as a characteristic of individuals that they might carry from setting to setting, or as a characteristic of economic behavior that can be measured against an economizing standard. The independent variable has repeatedly been conceptualized in social, that is, relative terms, and the measurement has always been consistent with this conceptualization.

It is of course true that the same measurements might have been used with other conceptualizations; specifically, in this case, with one that sees innovation as a characteristic of individuals or as an economizing decision relative to absolute financial goals. However, the substantive propositions coming out of the theory used here concern the relation of rank and innovation in a community. That is, they concern the structure of communities, not the person or the economy. This is why I believe that the study shows the innovator to be a product of his or her situation relative to other people in the community.

### *Implications of This Study for Research on Stratification*

As is frequently pointed out, scholars have been thinking about the nature of social stratification for a long time. Thus it is unlikely that my conclusions are new, except perhaps in the way they combine old ideas. This fact makes them easy to state.

By emphasizing the relative nature of stratification systems, the approach taken here 1) calls attention to the importance of defining the community of reference and 2) raises a caution about the use of absolute underlying metrics (like dollars of income) in the study of stratification. The appearance of the dip in the relationship of rank and innovation that is the manifestation of upper-middle-class conservatism in the particular context studied here only heightens the need to attend to the situation from the point of view of the local actors. I have tried to suggest, however, that this attention to the actor's situation does not demand total particularism in analysis of the local situation. More specifically, the focus is on the actor's position in his or her community of reference, not on the actor's attribution of meaning to events or situations.

By emphasizing the effect of uncertainty in situations of change, the approach taken here points to the different evaluations of change that people at different places in an established hierarchy are likely to have. In general it calls attention to the need to complement focus on position within the community of reference with the realization that people in different positions see and define different systems of stratification with different degrees of ambivalence and ambiguity (Cancian 1976a). The local conception of the stratification system is as subject to distortion, self-serving change, and multivocality as other cultural products.

### *Implications of This Study for Agricultural Development Policy*

This study has a broad range of implications for agricultural development policy. They follow from three elements of the findings: the relation of rank and innovation; the primacy of social-situational over personal characteristics as explanatory principles; and the importance of responses to uncertainty, rather than communication processes, in the early stages of the spread of innovations.

The old relationship between economic rank and adoption of innovations is broken into two parts in this study: an early and a later stage in the spread of an innovation are distinguished. The conservatism of wealthier farmers (specifically those just above the middle of the community ranking system) displayed in the first stage of the adoption process means that change programs may fail because these farmers see no point in experimenting with the new. The overall linear relationship between rank and adoption of new agricultural practices found by the end of the second stage suggests that, in the long run, resources—plain old financial ability—still influence who might adopt new practices.

The implications of the finding about rank and innovation in Stage 1 for any specific policy or program will of course depend in substantial part on the policy-maker's interpretation of the multiple other factors that will be relevant. Today, when many development specialists are wondering about ways to direct aid to small farmers, the implication that is most interesting to me is: failure of a program to be accepted by what may be considered "barely viable" small farmers does not mean that it will not succeed with still smaller farmers who, by many standards, are too poor to offer hope that they can produce enough to meet national goals for their own minimal material well-being. Programs extended to farmers as small as many who fell into the low-middle rank in this study may have to be organized differently from present programs, because, for example, administrative costs per farm unit become relatively greater as the size of the unit decreases. Clearly, the poorer the farmer, the more difficult it is to efficiently serve him or her with personnel paid on the scale of national and international technocrats.* Thus, to sum up this extention of the findings about the relation of rank and risk during Stage 1 of the adoption process, recognition that the small farmer is innovative in inclination and apparently open to programs appropriate in scale to his or her resources challenges the development planner to find a way to deliver aid in an economically viable manner. The most perplexing part of this challenge promises to be identifying something useful that can be done across the gross income disparities that separate planners, technicians, and peasants.

Various aspects of this study suggest that social situation rather than personal characteristics should be given primacy in explaining adoption and in designing programs to promote change.

*Cancian 1978 expands on this point.

The primacy of the social situation has broad implications. It means that policy-makers should attend to the social and economic relationships maintained by any system of production they seek to alter. It means that policy implementers should attend to the local social structure and the boundaries of the locally-defined community; and that programs aimed at individuals without attention to their social position will, at best, be less than optimally efficient in the delivery of aid.

Finally, the framework that made it possible to understand the unusual pattern found in this research includes heavy emphasis on the role of uncertainty in the early stages of the spread of an innovation. The importance of uncertainty in explaining the differences between Stage 1 and Stage 2 of the adoption process highlights the role of uncertainty in practical decision-making. Farmers dealing with innovation cannot be simple dispassionate economic maximizers. They seldom have enough information (Cancian 1979b, Cancian n.d.). On the other hand, this study suggests that they do not randomly grasp at alternatives. Rather, they seem to assess the implications of their decisions under uncertainty for their larger situation, that is for their future economic standing in their community. For a policy and program establishment designed, for good reason, to provide clear information that bears directly on the technical and economic issues, it may be difficult to absorb the clear implication that adoption seems to result, for good reason, from decisions made with partial information and considering social factors that have little to do with the technical and economic problems. And, more important still, these noneconomically-rational factors are clearly "rational" within the larger system and destined to remain central in any foreseeable practical transformation of the social and productive system. Thus, they should receive positive attention in development planning.

At the eleventh hour it seems worth noting where this study stands in relation to both the new wave of macro/structural studies of agriculture and the traditional micro/decision-making studies of agriculture. Adherents of dependency theory and other macro approaches to the relation of more and less developed countries point out that the traditional emphasis on personal characteristics and communication of information is part of an orientation that favors the developed nations. The validity of this criticism has been recognized by many policy makers who have used the criticized orientations; but the macro approach has no clear and useful implications for the economic development establishment that works at the local level. The present study, which falls between the traditional individualistic approach and the recently popular macro approach, does offer an alternative that is practical.

In general, this study suggests that agricultural development, insofar as it involves the spread of new farming practices, depends heavily on both the local stratification system and the less than fully informed economic decision making that dominates everyday life. The sooner program design explicitly attends to these realities, the sooner it will come closer to achieving its explicitly stated goals.

# *Appendixes*
# *References*

# *Appendix A*

# *Data and Tests for the Crosstabular Analysis*

This appendix has four parts: 1) rules for dividing the variables for crosstabular analysis using three different sets of cutting points for the independent variable; 2) the tables resulting from application of these rules to each available set of variables for each data set; 3) procedures used to standardize the Stage 1 and Stage 2 adoption rates; and 4) a table of standardized rates and test results based on them. Part 2) includes tables for all the cases displayed in Figure 8 except the last seven. For them comparable tables are found in Cancian 1967: 922.

### *Rules for Dividing Variables*

The independent variables were divided into quartiles (25/25/25/25) using the following rules to deal with lumpiness in the available data. Take the best approximation to a 50/50 division of the distribution. If there is a tie, make the lower half larger, that is, favor fifty-one cases below the cutting point and forty-nine above it over the reverse. Divide each of the halves in half. If there is a tie, make the middle rank larger. Cutting points for the 30/30/20/20 division were determined by using 60/40 in place of 50/50 plus the same subsequent rules.

To divide the independent variables into deciles the following rules were used. Start from the low end of the distribution. Stay as close to round decile cuts (10 percent, 20 percent, 30 percent) as possible. If tied in closeness to pure cuts, favor smaller deciles. No decile should include less than thirteen cases. Noniles, octiles, and septiles are acceptable outcomes if the data are lumpy; sextiles are not. If exceptions are made, note the reasons.

The dependent variable was divided as follows. Start with the first adopters or the highest point of the adoption index. Stage 1 is the best approximation to the first 25 percent of farmers. Favor the smaller number in case of a tie. Then take the best approximation to the second 25 percent with the same conditions.

If the resulting four-rank crosstabulations produced expected values of less than five for any of the cells in Stage 1, the variables were inspected for alternative cutting points that would increase the size of Stage 1. The cutting point for Taiwan 159 was changed on this basis.

### *The Crosstabulation Tables: A.1*

Each of the pages in this table (which appears at the end of the text of the appendix) shows tables for one independent variable for one case. The seven cases that have two independent variables are represented by two pages each. As noted above, figures for the seven cases taken from Cancian 1967 are not repeated here. Each page here displays the 25/25 (Q), the 30/20 (T), and the decile (D) distributions if the original data permit them to be calculated.

| | |
|---|---|
| 1) India 185 | 9) Philippines 177* |
| 2) India 412* | 10) Taiwan 159 |
| 3) Kenya 540 | 11) Taiwan 237 |
| 4) Mexico 108 | 12) U.S. 159* |
| 5) Mexico 193 | 13) U.S. 167* |
| 6) Pakistan 221 | 14) India 676* |
| 7) Pakistan 350 | 15) India 1,145* |
| 8) Philippines 153* | 16) U.S. 316 |

Cases marked with an asterisk have one page for each of two independent variables. All the area variables are in hectares. The various income and value variables are in local currencies as listed in Table 7.

### *Procedures Used for Standardizing Adoption Rates in Stage 1 and Stage 2*

Two kinds of transformations of the raw figures and percentages presented in the tables above may be used for testing the hypotheses. They enhance the comparability of results, and they affect the outcome of tests of Hypothesis B in some cases.

The first transformation concerns the adoption rate in Stage 2. Is it to be calculated by dividing the number of farmers adopting by the total number of farmers in the rank, or by the number who have not adopted in Stage 1? The latter alternative seems more appropriate—both because it concentrates on the people "available" to adopt in Stage 2, and because it "works against" Hypothesis B in situations where Hypothesis A has been confirmed, and thus avoids some essentially artifactual confirmations of Hypothesis B. Thus, in making the calculations reported below, the adoption rate for Stage 2 was calculated after eliminating the farmers who had already adopted in Stage 1. It should be noted that, given the previously stated rules for definition of stages, this convention implies that the mean adoption rate for Stage 2 will be 33 percent rather than the parallel figure of 25 percent for Stage 1.

Another kind of distortion of relative adoption rates from the stage to stage results from the fact that the stages themselves (as operationalized) are not perfect quartiles, because of lumpiness in the original data. This distortion is not systematic, as is the type discussed above.

Assuming that the stages as operationalized are fair estimates of the theoretical stages, the distortions introduced by both lumpiness in the data and the calculation rules stated above may be eliminated by converting the figures to a standard base. This procedure, used in my earlier study (Cancian 1967: 923), makes the sum of adoption rates for each stage equal to 100.

Gartrell, Wilkening, and Presser (1973: 397) make interpretations that suggest they are not aware of the essentially artifactual nature of the absolute sizes of standardized adoption rates for Stage 2 in relation to comparable rates for Stage 1. The hypothesis they discuss (B here, 6 in Cancian 1967) is clearly meant to imply that, from Stage 1 to Stage 2, the high-middle rank will gain in adoption rate relative to the low-middle rank. Given the conventions used in calculations, a confirmation of the hypothesis ($HM_2 - HM_1 > LM_2 - LM_1$) may take the form: $22 - 24 > 24 - 27$ (as it does in the case of Mexico 193, shown in Table A.2).

Table A.2 shows results of subjecting the data in Table A.1 to both transformations described above. The columns at the right in Table A.2 show the results of testing the hypotheses.

The test results with the "raw" column percentages in Table A.1 and the standardized figures in Table A.2 are, of course, always the same for Hypothesis A. For Hypothesis B the outcome is different in three of the forty-nine cases. The cases (Kenya 540, 25/25; Taiwan 237; and U.S. 167, area, 30/20) are marked in the Hypothesis B column of Table A.2. The outcomes which change with the changes in calculation procedures represent "close" cases in which there is no issue of substantial or statistically significant difference between the alternative outcomes. While these cases are not crucial to the outcome of the study, it would be useful to specify Hypothesis B so that decisions could be made on these three cases.

This can be done as follows. Hypothesis B states that high-middle-rank farmers will adopt relatively more in Stage 2; that is, relatively more in comparison with themselves in relation to low-middle-rank farmers in Stage 1. This idea has been specified as $HM_2 - HM_1 > LM_2 - LM_1$, where rank adoption rates are entered into the equation. The differences under discussion result from different methods of calculating adoption rates.

The hypothesis may also be specified (in terms of raw numbers) as follows: $HM_2/(HM_1 + HM_2) > LM_2/(LM_1 + LM_2)$ and $HM_2/(HM_2 + LM_2) > HM_1/(HM_1 + LM_1)$. The results from these specifications agree with the "standardized" figures in the Taiwan case and not in the Kenya and U.S. cases. This outcome, which makes the results on Hypothesis B consistent with results on Hypothesis A, is used in tabulation of the results.

TABLE A.1. *Full Crosstabulations for All Data Sets*

*1) India 185*

Area Q

| Count<br>Col % | 0.2-<br>1.3 | 1.4-<br>2.4 | 2.6-<br>5.9 | 6.1-<br>32.5 | Row<br>Total |
|---|---|---|---|---|---|
| Remainder | 12<br>66.7 | 13<br>48.1 | 24<br>63.2 | 12<br>27.3 | 61<br>48.0 |
| Stage 2 | 6<br>33.3 | 10<br>37.0 | 10<br>26.3 | 12<br>27.3 | 38<br>29.9 |
| Stage 1 | 0<br>0.0 | 4<br>14.8 | 4<br>10.5 | 20<br>45.5 | 28<br>22.0 |
| Column<br>Total | 18<br>14.2 | 27<br>21.3 | 38<br>29.9 | 44<br>34.6 | 127<br>100.0 |

Number of Missing Observations = 58

Area T

| Count<br>Col % | 0.2-<br>1.5 | 1.6-<br>3.0 | 3.2-<br>6.5 | 6.9-<br>32.5 | Row<br>Total |
|---|---|---|---|---|---|
| Remainder | 16<br>66.7 | 17<br>51.5 | 18<br>52.9 | 10<br>27.8 | 61<br>48.0 |
| Stage 2 | 7<br>29.2 | 11<br>33.3 | 10<br>29.4 | 10<br>27.8 | 38<br>29.9 |
| Stage 1 | 1<br>4.2 | 5<br>15.2 | 6<br>17.6 | 16<br>44.4 | 28<br>22.0 |
| Column<br>Total | 24<br>18.9 | 33<br>26.0 | 34<br>26.8 | 36<br>28.3 | 127<br>100.0 |

Number of Missing Observations = 58

Area D

| Count<br>Col % | 0.2-<br>0.5 | 0.6-<br>1.0 | 1.2-<br>1.5 | 1.6-<br>1.9 | 2.0-<br>2.4 | 2.6-<br>3.0 | 3.2-<br>4.4 | 4.5-<br>6.5 | 6.9-<br>11.7 | 12.1-<br>32.5 | Row<br>Total |
|---|---|---|---|---|---|---|---|---|---|---|---|
| Remainder | 2<br>66.7 | 5<br>62.5 | 9<br>69.2 | 4<br>50.0 | 5<br>38.5 | 8<br>66.7 | 11<br>64.7 | 7<br>41.2 | 4<br>26.7 | 6<br>28.6 | 61<br>48.0 |
| Stage 2 | 1<br>33.3 | 3<br>37.5 | 3<br>23.1 | 1<br>12.5 | 8<br>61.5 | 2<br>16.7 | 5<br>29.4 | 5<br>29.4 | 7<br>46.7 | 3<br>14.3 | 38<br>29.9 |
| Stage 1 | 0<br>0.0 | 0<br>0.0 | 1<br>7.7 | 3<br>37.5 | 0<br>0.0 | 2<br>16.7 | 1<br>5.9 | 5<br>29.4 | 4<br>26.7 | 12<br>57.1 | 28<br>22.0 |
| Column<br>Total | 3<br>2.4 | 8<br>6.3 | 13<br>10.2 | 8<br>6.3 | 13<br>10.2 | 12<br>9.4 | 17<br>13.4 | 17<br>13.4 | 15<br>11.8 | 21<br>16.5 | 127<br>100.0 |

Number of Missing Observations = 58

*2) India 412*

Area Q

| Count<br>Col % | .2-<br>.5 | .7-<br>.9 | 1.2-<br>1.6 | 1.9-<br>8.1 | Row<br>Total |
|---|---|---|---|---|---|
| Remainder | 62<br>62.0 | 56<br>45.2 | 47<br>50.5 | 38<br>40.0 | 203<br>49.3 |
| Stage 2 | 31<br>31.0 | 36<br>29.0 | 20<br>21.5 | 29<br>30.5 | 116<br>28.2 |
| Stage 1 | 7<br>7.0 | 32<br>25.8 | 26<br>28.0 | 28<br>29.5 | 93<br>22.6 |
| Column<br>Total | 100<br>24.3 | 124<br>30.1 | 93<br>22.6 | 95<br>23.1 | 412<br>100.0 |

Area T

| Count<br>Col % | .2-<br>.5 | .7-<br>1.2 | 1.4-<br>1.9 | 2.1-<br>8.1 | Row<br>Total |
|---|---|---|---|---|---|
| Remainder | 62<br>62.0 | 79<br>48.5 | 38<br>48.7 | 24<br>33.8 | 203<br>49.3 |
| Stage 2 | 31<br>31.0 | 42<br>25.8 | 21<br>26.9 | 22<br>31.0 | 116<br>28.2 |
| Stage 1 | 7<br>7.0 | 42<br>25.8 | 19<br>24.4 | 25<br>35.2 | 93<br>22.6 |
| Column<br>Total | 100<br>24.3 | 163<br>39.6 | 78<br>18.9 | 71<br>17.2 | 412<br>100.0 |

Area D

| Count<br>Col % | .2 only | .5 only | .7 only | .9 only | 1.2 only | 1.4 only | 1.6-<br>1.9 | 2.1-<br>2.8 | 3.0-<br>8.1 | Row<br>Total |
|---|---|---|---|---|---|---|---|---|---|---|
| Remainder | 15<br>68.2 | 47<br>60.3 | 32<br>50.0 | 24<br>40.0 | 23<br>59.0 | 14<br>43.8 | 24<br>52.2 | 18<br>41.9 | 6<br>21.4 | 203<br>49.3 |
| Stage 2 | 3<br>13.6 | 28<br>35.9 | 18<br>28.1 | 18<br>30.0 | 6<br>15.4 | 8<br>25.0 | 13<br>28.3 | 13<br>30.2 | 9<br>32.1 | 116<br>28.2 |
| Stage 1 | 4<br>18.2 | 3<br>3.8 | 14<br>21.9 | 18<br>30.0 | 10<br>25.6 | 10<br>31.3 | 9<br>19.6 | 12<br>27.9 | 13<br>46.4 | 93<br>22.6 |
| Column<br>Total | 22<br>5.3 | 78<br>18.9 | 64<br>15.5 | 60<br>14.6 | 39<br>9.5 | 32<br>7.8 | 46<br>11.2 | 43<br>10.4 | 28<br>6.8 | 412<br>100.0 |

TABLE A.1. *Continued*

*2) India 412*

Value Product Q

| Count<br>Col % | 10-<br>100 | 110-<br>220 | 230-<br>440 | 450-<br>3270 | Row<br>Total |
|---|---|---|---|---|---|
| Remainder | 65<br>63.7 | 63<br>60.6 | 43<br>41.3 | 32<br>31.4 | 203<br>49.3 |
| Stage 2 | 25<br>24.5 | 31<br>29.8 | 33<br>31.7 | 27<br>26.5 | 116<br>28.2 |
| Stage 1 | 12<br>11.8 | 10<br>9.6 | 28<br>26.9 | 43<br>42.2 | 93<br>22.6 |
| Column<br>Total | 102<br>24.8 | 104<br>25.2 | 104<br>25.2 | 102<br>24.8 | 412<br>100.0 |

Value Product T

| Count<br>Col % | 10-<br>120 | 130-<br>300 | 310-<br>510 | 520-<br>3270 | Row<br>Total |
|---|---|---|---|---|---|
| Remainder | 78<br>64.5 | 69<br>53.5 | 28<br>35.4 | 28<br>33.7 | 203<br>49.3 |
| Stage 2 | 29<br>24.0 | 42<br>32.6 | 26<br>32.9 | 19<br>22.9 | 116<br>28.2 |
| Stage 1 | 14<br>11.6 | 18<br>14.0 | 25<br>31.6 | 36<br>43.4 | 93<br>22.6 |
| Column<br>Total | 121<br>29.4 | 129<br>31.3 | 79<br>19.2 | 83<br>20.1 | 412<br>100.0 |

Value Product D

| Count<br>Col % | 10-<br>50 | 60-<br>80 | 90-<br>120 | 130-<br>160 | 170-<br>220 | 230-<br>300 | 310-<br>370 | 380-<br>510 | 520-<br>760 | 770-<br>3270 | Row<br>Total |
|---|---|---|---|---|---|---|---|---|---|---|---|
| Remainder | 29<br>76.3 | 24<br>60.0 | 25<br>58.1 | 26<br>61.9 | 24<br>55.8 | 19<br>43.2 | 15<br>37.5 | 13<br>33.3 | 18<br>42.9 | 10<br>24.4 | 203<br>49.3 |
| Stage 2 | 8<br>21.1 | 12<br>30.0 | 9<br>20.9 | 11<br>26.2 | 16<br>37.2 | 15<br>34.1 | 13<br>32.5 | 13<br>33.3 | 7<br>16.7 | 12<br>29.3 | 116<br>28.2 |
| Stage 1 | 1<br>2.6 | 4<br>10.0 | 9<br>20.9 | 5<br>11.9 | 3<br>7.0 | 10<br>22.7 | 12<br>30.0 | 13<br>33.3 | 17<br>40.5 | 19<br>46.3 | 93<br>22.6 |
| Column<br>Total | 38<br>9.2 | 40<br>9.7 | 43<br>10.4 | 42<br>10.2 | 43<br>10.4 | 44<br>10.7 | 40<br>9.7 | 39<br>9.5 | 42<br>10.2 | 41<br>10.0 | 412<br>100.0 |

**TABLE A.1.** *Continued*

*3) Kenya 540*

Wealth Index Q

| Count<br>Row %<br>Col %<br>Total | L | LM | HM | H | Row Total |
|---|---|---|---|---|---|
| Remainder | 109<br>60.2 | 27<br>45.8 | 49<br>41.2 | 69<br>42.3 | 254<br>48.7 |
| Stage 2 | 37<br>20.4 | 17<br>28.8 | 37<br>31.1 | 43<br>26.4 | 134<br>48.7 |
| Stage 1 | 35<br>19.3 | 15<br>25.4 | 33<br>27.7 | 51<br>31.3 | 134<br>25.7 |
| Column Total | 181<br>34.7 | 59<br>11.3 | 119<br>22.8 | 163<br>31.2 | 522<br>100.0 |

Number of Missing Observations = 18

Wealth Index T

| Count<br>Row %<br>Col %<br>Total% | L | LM | HM | H | Row Total |
|---|---|---|---|---|---|
| Remainder | 109<br>60.2 | 51<br>45.1 | 52<br>41.9 | 42<br>40.0 | 254<br>48.7 |
| Stage 2 | 37<br>20.4 | 29<br>25.7 | 39<br>31.5 | 29<br>27.9 | 134<br>25.7 |
| Stage 1 | 35<br>19.3 | 33<br>29.2 | 33<br>26.6 | 33<br>31.7 | 134<br>25.7 |
| Column Total | 181<br>34.7 | 113<br>21.6 | 124<br>23.8 | 104<br>19.9 | 522<br>100.0 |

Number of Missing Observations = 18

Wealth Index D

| Count<br>Col % | 1. | 2. | 3. | 4. | 5. | 6. | 7. | 8. | Row Total |
|---|---|---|---|---|---|---|---|---|---|
| Remainder | 27<br>81.8 | 82<br>55.4 | 27<br>45.8 | 24<br>44.4 | 25<br>38.5 | 27<br>45.8 | 24<br>42.9 | 18<br>37.5 | 254<br>48.7 |
| Stage 2 | 5<br>15.2 | 32<br>21.6 | 17<br>28.8 | 12<br>22.2 | 25<br>38.5 | 14<br>23.7 | 16<br>28.6 | 13<br>27.1 | 134<br>25.7 |
| Stage 1 | 1<br>3.0 | 34<br>23.0 | 15<br>25.4 | 18<br>33.3 | 15<br>23.1 | 18<br>30.5 | 16<br>28.6 | 17<br>35.4 | 134<br>25.7 |
| Column Total | 33<br>6.3 | 148<br>28.4 | 59<br>11.3 | 54<br>10.3 | 65<br>12.5 | 59<br>11.3 | 56<br>10.7 | 48<br>9.2 | 522<br>100.0 |

Number of Missing Observations = 18

TABLE A.1. *Continued*

*4) Mexico 108*

Area Q

| Count<br>Col % | .75 only | 1.5 only | 2.25 only | 3.0<br>6.0 | Row<br>Total |
|---|---|---|---|---|---|
| Remainder | 16<br>88.9 | 23<br>71.9 | 19<br>61.3 | 11<br>40.7 | 69<br>63.9 |
| Stage 2 | 0<br>0.0 | 4<br>12.5 | 11<br>35.5 | 5<br>18.5 | 23<br>18.5 |
| Stage 1 | 2<br>11.1 | 5<br>15.6 | 1<br>3.2 | 11<br>40.7 | 19<br>17.6 |
| Column<br>Total | 18<br>16.7 | 32<br>29.6 | 31<br>28.7 | 27<br>25.0 | 108<br>100.0 |

Area D*

| Count<br>Col % | .75 | 1.5 | 2.25 | 30 | 3.75 | 4.5 | 5.25 | 60 | Row<br>Total |
|---|---|---|---|---|---|---|---|---|---|
| Remainder | 16<br>88.9 | 23<br>71.9 | 19<br>61.3 | 7<br>63.6 | 3<br>75.0 | 0<br>0.0 | 0<br>0.0 | 1<br>100.0 | 69<br>63.9 |
| Stage 2 | 0<br>0.0 | 4<br>12.5 | 11<br>35.5 | 2<br>18.2 | 0<br>0.0 | 2<br>25.0 | 1<br>33.3 | 0<br>0.0 | 20<br>18.5 |
| Stage 1 | 2<br>11.1 | 5<br>15.6 | 1<br>3.2 | 2<br>18.2 | 1<br>25.0 | 6<br>75.0 | 2<br>66.7 | 0<br>0.0 | 19<br>17.6 |
| Column<br>Total | 18<br>16.7 | 32<br>29.6 | 31<br>28.7 | 11<br>10.2 | 4<br>3.7 | 8<br>7.4 | 3<br>2.8 | 1<br>0.9 | 108<br>100.0 |

*Fully detailed distribution. Not lumped by rules for deciles. See Appendix C.

TABLE A.1. *Continued*

*5) Mexico 193*

Area Q

| Count<br>Col % | .75 only | 1.5 and<br>2.75 | 3.0 and<br>3.75 | 4.5-<br>6.0 | Row<br>Total |
|---|---|---|---|---|---|
| Remainder | 23<br>71.9 | 43<br>55.8 | 28<br>59.6 | 15<br>40.5 | 109<br>56.5 |
| Stage 2 | 5<br>15.6 | 12<br>15.6 | 7<br>14.9 | 7<br>18.9 | 31<br>16.1 |
| Stage 1 | 4<br>12.5 | 22<br>28.6 | 12<br>25.5 | 15<br>40.5 | 53<br>27.5 |
| Column<br>Total | 32<br>16.6 | 77<br>39.9 | 47<br>24.4 | 37<br>19.2 | 193<br>100.0 |

Area D*

| Count<br>Col % | .75 | 1.5 | 2.25 | 3.0 | 3.75 | 4.5 | 5.25 | 60 | Row<br>Total |
|---|---|---|---|---|---|---|---|---|---|
| Remainder | 23<br>71.9 | 32<br>66.7 | 11<br>37.9 | 20<br>60.6 | 8<br>57.1 | 10<br>45.5 | 0<br>0.0 | 5<br>35.7 | 109<br>56.5 |
| Stage 2 | 5<br>15.6 | 4<br>8.3 | 8<br>27.6 | 5<br>15.2 | 2<br>14.3 | 5<br>22.7 | 0<br>0.0 | 2<br>14.3 | 31<br>16.1 |
| Stage 1 | 4<br>12.5 | 12<br>25.0 | 10<br>34.5 | 8<br>24.2 | 4<br>28.6 | 7<br>31.8 | 1<br>100.0 | 7<br>50.0 | 53<br>27.5 |
| Column<br>Total | 32<br>16.6 | 48<br>24.9 | 29<br>15.0 | 33<br>17.1 | 14<br>7.3 | 22<br>11.4 | 1<br>0.5 | 14<br>7.3 | 193<br>100.0 |

*Fully detailed distribution. Not lumped by rules for Deciles. See Appendix C.

TABLE A.1. *Continued*
*6) Pakistan 221*

Area Q

| Count<br>Col % | .05-<br>.28 | .30-<br>.48 | .51-<br>.86 | .91-<br>5.06 | Row<br>Total |
|---|---|---|---|---|---|
| Remainder | 19<br>41.3 | 18<br>30.5 | 16<br>29.1 | 18<br>29.5 | 71<br>32.1 |
| Stage 2 | 16<br>34.8 | 23<br>39.0 | 21<br>38.2 | 15<br>24.6 | 75<br>33.9 |
| Stage 1 | 11<br>23.9 | 18<br>30.5 | 18<br>32.7 | 28<br>45.9 | 74<br>33.9 |
| Column<br>Total | 46<br>20.8 | 59<br>26.7 | 55<br>24.9 | 61<br>27.6 | 221<br>100.0 |

Area T

| Count<br>Col % | .05-<br>.30 | .35-<br>.56 | .61-<br>1.01 | 1.06-<br>5.06 | Row<br>Total |
|---|---|---|---|---|---|
| Remainder | 26<br>41.3 | 20<br>29.9 | 16<br>32.0 | 9<br>22.0 | 71<br>32.1 |
| Stage 2 | 22<br>34.9 | 24<br>35.8 | 21<br>42.0 | 8<br>19.5 | 75<br>33.9 |
| Stage 1 | 15<br>23.8 | 23<br>34.3 | 13<br>26.0 | 24<br>58.5 | 75<br>33.9 |
| Column<br>Total | 63<br>28.5 | 67<br>30.3 | 50<br>22.6 | 41<br>18.6 | 221<br>100.0 |

Area D

| Count<br>Col % | .05-<br>.19 | .20-<br>.25 | .28-<br>.30 | .35-<br>.40 | .46-<br>.51 | .56-<br>.66 | .71-<br>.76 | .81-<br>1.01 | 1.06-<br>1.37 | 1.56-<br>5.06 | Row<br>Total |
|---|---|---|---|---|---|---|---|---|---|---|---|
| Remainder | 11<br>50.0 | 8<br>34.8 | 7<br>38.9 | 11<br>32.4 | 8<br>26.7 | 3<br>23.1 | 4<br>30.8 | 10<br>37.0 | 4<br>22.2 | 5<br>21.7 | 71<br>32.1 |
| Stage 2 | 9<br>36.4 | 8<br>34.8 | 6<br>33.3 | 13<br>38.2 | 10<br>33.3 | 7<br>53.8 | 5<br>38.5 | 10<br>37.0 | 4<br>22.2 | 4<br>17.4 | 75<br>33.9 |
| Stage 1 | 3<br>13.6 | 7<br>30.4 | 5<br>27.8 | 10<br>29.4 | 12<br>40.0 | 3<br>23.1 | 4<br>30.8 | 7<br>25.9 | 10<br>55.6 | 14<br>60.9 | 75<br>33.9 |
| Column<br>Total | 22<br>10.0 | 23<br>10.4 | 18<br>8.1 | 34<br>15.4 | 30<br>13.6 | 13<br>5.9 | 13<br>5.9 | 27<br>12.2 | 18<br>8.1 | 23<br>10.4 | 221<br>100.0 |

TABLE A.1. *Continued*

*7) Pakistan 350* [a]

Area Q

| Count<br>Col % | .2-<br>1.0 | 1.1-<br>1.7 | 1.8-<br>2.8 | 3.0-<br>132.7 | Row<br>Total |
|---|---|---|---|---|---|
| Remainder | 17<br>37.0 | 15<br>31.3 | 22<br>28.6 | 34<br>19.0 | 88<br>25.1 |
| Stage 2 | 27<br>58.7 | 27<br>56.3 | 50<br>64.9 | 102<br>57.0 | 206<br>58.9 |
| Stage 1 | 2<br>4.3 | 6<br>12.5 | 5<br>6.5 | 43<br>24.0 | 56<br>16.0 |
| Column<br>Total | 46<br>13.1 | 48<br>13.7 | 77<br>22.0 | 179<br>51.1 | 350<br>100.0 |

Area T

| Count<br>Col % | .2-<br>1.1 | 1.2-<br>1.9 | 2.0-<br>3.2 | 3.4-<br>132.7 | Row<br>Total |
|---|---|---|---|---|---|
| Remainder | 17<br>36.2 | 19<br>32.2 | 23<br>28.8 | 29<br>17.7 | 88<br>25.1 |
| Stage 2 | 28<br>59.6 | 33<br>55.9 | 51<br>63.8 | 94<br>57.3 | 206<br>58.9 |
| Stage 1 | 2<br>4.3 | 7<br>11.9 | 6<br>7.5 | 41<br>25.0 | 56<br>16.0 |
| Column<br>Total | 47<br>13.4 | 59<br>16.9 | 80<br>22.9 | 164<br>46.9 | 350<br>100.0 |

Area D

| Count<br>Col % | .2-<br>.7 | .8-<br>.9 | 1-<br>1.1 | 1.2-<br>1.4 | 1.5-<br>1.7 | 1.8-<br>1.9 | 2.0-<br>2.4 | 2.5-<br>3.2 | 3.4-<br>4.9 | 5.0-<br>132.7 | Row<br>Total |
|---|---|---|---|---|---|---|---|---|---|---|---|
| Remainder | 8<br>42.1 | 8<br>50.0 | 1<br>8.3 | 9<br>33.3 | 6<br>30.0 | 4<br>33.3 | 14<br>28.6 | 9<br>29.0 | 14<br>29.2 | 15<br>12.9 | 88<br>25.1 |
| Stage 2 | 9<br>47.4 | 8<br>50.0 | 11<br>91.7 | 15<br>55.6 | 11<br>55.0 | 7<br>58.3 | 33<br>67.3 | 18<br>58.1 | 31<br>64.6 | 63<br>54.3 | 206<br>58.9 |
| Stage 1 | 2<br>10.5 | 0<br>0.0 | 0<br>0.0 | 3<br>11.1 | 3<br>15.0 | 1<br>8.3 | 2<br>4.1 | 4<br>12.9 | 3<br>6.3 | 38<br>32.8 | 56<br>16.0 |
| Column<br>Total | 19<br>5.4 | 16<br>4.6 | 12<br>3.4 | 27<br>7.7 | 20<br>5.7 | 12<br>3.4 | 49<br>14.0 | 31<br>8.9 | 48<br>13.7 | 116<br>33.1 | 350<br>100.0 |

[a]These figures represent the actual cases reported by Lowdermilk. He stratified by farm size and took a non-proportional, random sample (see Table 3). The cutting points used here were determined on the basis of the reconstituted "representative" distribution of farm size.

TABLE A.1. *Continued*

*8) Philippines 153*

Area Q

| Count<br>Col % | .1-<br>.6 | .7-<br>1.1 | 1.2-<br>1.9 | 2.0-<br>21.3 | Row<br>Total |
|---|---|---|---|---|---|
| Remainder | 22<br>61.1 | 28<br>65.1 | 24<br>61.5 | 16<br>45.7 | 90<br>58.8 |
| Stage 2 | 8<br>22.2 | 10<br>23.3 | 7<br>17.9 | 6<br>17.1 | 31<br>20.3 |
| Stage 1 | 6<br>16.7 | 5<br>11.6 | 8<br>20.5 | 13<br>37.1 | 32<br>20.9 |
| Column<br>Total | 36<br>23.5 | 43<br>28.1 | 39<br>25.5 | 35<br>22.9 | 153<br>100.0 |

Area T

| Count<br>Col % | .1-<br>.7 | .8-<br>1.3 | 1.4-<br>1.9 | 2.0-<br>21.3 | Row<br>Total |
|---|---|---|---|---|---|
| Remainder | 26<br>61.9 | 30<br>60.0 | 18<br>69.2 | 16<br>45.7 | 90<br>58.8 |
| Stage 2 | 10<br>23.8 | 11<br>22.0 | 4<br>15.4 | 6<br>17.1 | 31<br>20.3 |
| Stage 1 | 6<br>14.3 | 9<br>18.0 | 4<br>15.4 | 13<br>37.1 | 32<br>20.9 |
| Column<br>Total | 42<br>27.5 | 50<br>32.7 | 26<br>17.0 | 35<br>22.9 | 153<br>100.0 |

Area D

| Count<br>Col % | .1-<br>.3 | .4-<br>.6 | .7-<br>.8 | .9-<br>1.0 | 1.1-<br>1.2 | 1.3-<br>1.5 | 1.6-<br>1.9 | 2.0-<br>2.5 | 2.6-<br>21.3 | Row<br>Total |
|---|---|---|---|---|---|---|---|---|---|---|
| Remainder | 8<br>44.4 | 14<br>77.8 | 10<br>62.5 | 14<br>66.7 | 8<br>61.5 | 9<br>52.9 | 11<br>73.3 | 9<br>60.0 | 7<br>35.0 | 90<br>58.8 |
| Stage 2 | 7<br>38.9 | 1<br>5.6 | 4<br>25.0 | 5<br>23.8 | 2<br>15.4 | 4<br>23.5 | 2<br>13.3 | 1<br>6.7 | 5<br>25.0 | 31<br>20.3 |
| Stage 1 | 3<br>16.7 | 3<br>16.7 | 2<br>12.5 | 2<br>9.5 | 3<br>23.1 | 4<br>23.5 | 2<br>13.3 | 5<br>33.3 | 8<br>40.0 | 32<br>20.9 |
| Column<br>Total | 18<br>11.8 | 18<br>11.8 | 16<br>10.5 | 21<br>13.7 | 13<br>8.5 | 17<br>11.1 | 15<br>9.8 | 15<br>9.8 | 20<br>13.1 | 153<br>100.0 |

TABLE A.1. *Continued*

*8) Philippines 153*

Product Sold Q

| Count<br>Col % | 0-<br>54 | 55-<br>171 | 172-<br>400 | 401-<br>4184 | Row<br>Total |
|---|---|---|---|---|---|
| Remainder | 26<br>70.3 | 21<br>53.8 | 24<br>66.7 | 19<br>46.3 | 90<br>58.8 |
| Stage 2 | 8<br>21.6 | 10<br>25.6 | 4<br>11.1 | 9<br>22.0 | 31<br>20.3 |
| Stage 1 | 3<br>8.1 | 8<br>20.5 | 8<br>22.2 | 13<br>31.7 | 32<br>20.9 |
| Column<br>Total | 37<br>24.2 | 39<br>25.5 | 36<br>23.5 | 41<br>26.8 | 153<br>100.0 |

Product Sold T

| Count<br>Col % | 0-<br>68 | 69-<br>250 | 251-<br>531 | 532-<br>4184 | Row<br>Total |
|---|---|---|---|---|---|
| Remainder | 31<br>67.4 | 28<br>60.9 | 18<br>60.0 | 13<br>41.9 | 90<br>58.8 |
| Stage 2 | 11<br>23.9 | 7<br>15.2 | 5<br>16.7 | 8<br>25.8 | 31<br>20.3 |
| Stage 1 | 4<br>8.7 | 11<br>23.9 | 7<br>23.3 | 10<br>32.3 | 32<br>20.9 |
| Column<br>Total | 46<br>30.1 | 46<br>30.1 | 30<br>19.6 | 31<br>20.3 | 153<br>100.0 |

Product Sold D

| Count<br>Col % | 0 only | 4-<br>46 | 47-<br>68 | 69-<br>126 | 127-<br>171 | 172-<br>250 | 251-<br>360 | 361-<br>531 | 532-<br>800 | 801-<br>4184 | Row<br>Total |
|---|---|---|---|---|---|---|---|---|---|---|---|
| Remainder | 13<br>68.4 | 9<br>69.2 | 9<br>64.3 | 7<br>43.8 | 9<br>64.3 | 12<br>75.0 | 11<br>73.3 | 7<br>46.7 | 7<br>46.7 | 6<br>37.5 | 90<br>58.8 |
| Stage 2 | 4<br>21.1 | 3<br>23.1 | 4<br>28.6 | 5<br>31.3 | 2<br>14.3 | 0<br>0.0 | 3<br>20.0 | 2<br>13.3 | 5<br>33.3 | 3<br>18.8 | 31<br>20.3 |
| Stage 1 | 2<br>10.5 | 1<br>7.7 | 1<br>7.1 | 4<br>25.0 | 3<br>21.4 | 4<br>25.0 | 1<br>6.7 | 6<br>40.0 | 3<br>20.0 | 7<br>43.8 | 32<br>20.9 |
| Column<br>Total | 19<br>12.4 | 13<br>8.5 | 14<br>9.2 | 16<br>10.5 | 14<br>9.2 | 16<br>10.5 | 15<br>9.8 | 15<br>9.8 | 15<br>9.8 | 16<br>10.5 | 153<br>100.0 |

TABLE A.1. *Continued*

*9) Philippines 177*

Area Q

| Count<br>Col % | .2-<br>.6 | .7-<br>1.0 | 1.1-<br>2.0 | 2.1-<br>35.0 | Row<br>Total |
|---|---|---|---|---|---|
| Remainder | 21<br>42.0 | 17<br>40.5 | 17<br>32.7 | 9<br>27.3 | 64<br>36.2 |
| Stage 2 | 16<br>32.0 | 10<br>23.8 | 17<br>32.7 | 12<br>36.4 | 55<br>31.1 |
| Stage 1 | 13<br>26.0 | 15<br>35.7 | 18<br>34.6 | 12<br>36.4 | 58<br>32.8 |
| Column<br>Total | 50<br>28.2 | 42<br>23.7 | 52<br>29.4 | 33<br>18.6 | 177<br>100.0 |

Area T

| Count<br>Col % | .2-<br>.7 | .8-<br>1.4 | 1.5-<br>2.0 | 2.1-<br>35.0 | Row<br>Total |
|---|---|---|---|---|---|
| Remainder | 21<br>41.2 | 23<br>39.7 | 11<br>31.4 | 9<br>27.3 | 64<br>36.2 |
| Stage 2 | 17<br>33.3 | 14<br>24.1 | 12<br>34.3 | 12<br>36.4 | 55<br>31.1 |
| Stage 1 | 13<br>25.5 | 21<br>36.2 | 12<br>34.3 | 12<br>36.4 | 58<br>32.8 |
| Column<br>Total | 51<br>28.8 | 58<br>32.8 | 35<br>19.8 | 33<br>18.6 | 177<br>100.0 |

Area D

| Count<br>Col % | .2-<br>.4 | .5 only | .6 only | .7-<br>.9 | 1.0 only | 1.1-<br>1.4 | 1.5-<br>1.8 | 1.9-<br>2.8 | 2.1-<br>3.0 | 3.1-<br>35.0 | Row<br>Total |
|---|---|---|---|---|---|---|---|---|---|---|---|
| Remainder | 8<br>40.0 | 8<br>47.1 | 5<br>38.5 | 3<br>23.1 | 14<br>48.3 | 6<br>35.3 | 4<br>26.7 | 7<br>35.0 | 6<br>28.6 | 3<br>25.0 | 64<br>36.2 |
| Stage 2 | 7<br>35.0 | 5<br>29.4 | 4<br>30.8 | 4<br>30.8 | 6<br>20.7 | 5<br>29.4 | 7<br>46.7 | 5<br>25.0 | 7<br>33.3 | 5<br>41.7 | 55<br>31.1 |
| Stage 1 | 5<br>25.0 | 4<br>23.5 | 4<br>30.8 | 6<br>46.2 | 9<br>31.0 | 6<br>35.3 | 4<br>26.7 | 8<br>40.0 | 8<br>38.1 | 4<br>33.3 | 58<br>32.8 |
| Column<br>Total | 20<br>11.3 | 17<br>9.6 | 13<br>7.3 | 13<br>7.3 | 29<br>16.4 | 17<br>9.6 | 15<br>8.5 | 20<br>11.3 | 21<br>11.9 | 12<br>6.8 | 177<br>100.0 |

TABLE A.1. *Continued*

*9) Philippines 177*

Product Sold Q

| Count<br>Col % | None | 4-<br>18 | 19-<br>150 | 151-<br>1000 | Row<br>Total |
|---|---|---|---|---|---|
| Remainder | 38<br>56.7 | 8<br>36.4 | 8<br>18.2 | 10<br>22.7 | 64<br>36.2 |
| Stage 2 | 17<br>25.4 | 8<br>36.4 | 17<br>38.6 | 13<br>29.5 | 55<br>31.1 |
| Stage 1 | 12<br>17.9 | 6<br>27.3 | 19<br>43.2 | 21<br>47.7 | 58<br>32.8 |
| Column<br>Total | 67<br>37.9 | 22<br>12.4 | 44<br>24.9 | 44<br>24.9 | 177<br>100.0 |

Product Sold T

| Count<br>Col % | None | 2-<br>40 | 41-<br>172 | 190-<br>1000 | Row<br>Total |
|---|---|---|---|---|---|
| Remainder | 38<br>56.7 | 9<br>23.1 | 10<br>27.8 | 7<br>20.0 | 64<br>36.2 |
| Stage 2 | 17<br>25.4 | 14<br>35.9 | 13<br>36.1 | 11<br>31.4 | 55<br>31.1 |
| Stage 1 | 12<br>17.9 | 16<br>41.0 | 13<br>36.1 | 17<br>48.6 | 58<br>32.8 |
| Column<br>Total | 67<br>37.9 | 39<br>22.0 | 36<br>20.3 | 35<br>19.8 | 177<br>100.0 |

Product Sold D

| Count<br>Col % | None | 4-<br>18 | 19-<br>40 | 41-<br>90 | 91-<br>172 | 173-<br>288 | 300-<br>1000 | Row<br>Total |
|---|---|---|---|---|---|---|---|---|
| Remainder | 38<br>56.7 | 8<br>36.4 | 1<br>5.9 | 4<br>22.2 | 6<br>33.3 | 5<br>29.4 | 2<br>11.1 | 64<br>36.2 |
| Stage 2 | 17<br>25.4 | 8<br>36.4 | 6<br>35.3 | 8<br>44.4 | 5<br>27.8 | 7<br>41.2 | 4<br>22.2 | 55<br>31.1 |
| Stage 1 | 12<br>17.9 | 6<br>27.3 | 10<br>58.8 | 6<br>33.3 | 7<br>38.9 | 5<br>29.4 | 12<br>66.7 | 58<br>32.8 |
| Column<br>Total | 67<br>37.9 | 22<br>12.4 | 17<br>9.6 | 18<br>10.2 | 18<br>10.2 | 17<br>9.6 | 18<br>10.2 | 177<br>100.0 |

TABLE A.1. *Continued*

*11) Taiwan 159*

Area T

| Count<br>Col % | LT 0.5 | 0.5-0.9 | 1.0-1.4 | 1.5 and over | Row Total |
|---|---|---|---|---|---|
| Remainder | 20<br>39.2 | 12<br>22.2 | 3<br>10.7 | 5<br>19.2 | 40<br>25.2 |
| Stage 2 | 8<br>15.7 | 9<br>16.7 | 9<br>32.1 | 8<br>30.8 | 34<br>21.4 |
| Stage 1 | 23<br>45.1 | 33<br>61.1 | 16<br>57.1 | 13<br>50.0 | 85<br>53.5 |
| Column Total | 51<br>32.1 | 54<br>34.0 | 28<br>17.6 | 26<br>16.4 | 159<br>100.0 |

*11) Taiwan 237*

Area Q

| Count<br>Col % | 0-<br>0.5 | 0.5-<br>0.9 | 1.0-<br>1.4 | 1.5 and over | Row Total |
|---|---|---|---|---|---|
| Remainder | 40<br>70.2 | 42<br>51.9 | 34<br>57.6 | 17<br>42.5 | 133<br>56.1 |
| Stage 2 | 10<br>17.5 | 23<br>28.4 | 15<br>25.4 | 17<br>42.5 | 65<br>27.4 |
| Stage 1 | 7<br>12.3 | 16<br>19.8 | 10<br>16.9 | 6<br>15.0 | 39<br>16.5 |
| Column Total | 57<br>24.1 | 81<br>34.2 | 59<br>24.9 | 40<br>16.9 | 237<br>100.0 |

TABLE A.1. *Continued*

*12) U.S. 159*

Area Q

| Count<br>Col % | 2-<br>87 | 88-<br>139 | 140-<br>231 | 232-<br>1397 | Row<br>Total |
|---|---|---|---|---|---|
| Remainder | 30<br>76.9 | 20<br>50.0 | 10<br>25.0 | 5<br>12.5 | 65<br>40.9 |
| Stage 2 | 5<br>12.8 | 13<br>32.5 | 13<br>32.5 | 10<br>25.0 | 41<br>25.8 |
| Stage 1 | 4<br>10.3 | 7<br>17.5 | 17<br>42.5 | 25<br>62.5 | 53<br>33.3 |
| Column<br>Total | 39<br>24.5 | 40<br>25.2 | 40<br>25.2 | 40<br>25.2 | 159<br>100.0 |

Area T

| Count<br>Col % | 2-<br>100 | 101-<br>180 | 181-<br>270 | 271-<br>1397 | Row<br>Total |
|---|---|---|---|---|---|
| Remainder | 36<br>73.5 | 18<br>40.0 | 9<br>27.3 | 2<br>6.3 | 65<br>40.9 |
| Stage 2 | 9<br>18.4 | 13<br>28.9 | 10<br>30.3 | 9<br>28.1 | 41<br>25.8 |
| Stage 1 | 4<br>8.2 | 14<br>31.1 | 14<br>42.4 | 21<br>65.6 | 53<br>33.3 |
| Column<br>Total | 49<br>30.8 | 45<br>28.3 | 33<br>20.8 | 32<br>20.1 | 159<br>100.0 |

Area D

| Count<br>Col % | 2-<br>60 | 61-<br>80 | 81-<br>97 | 98-<br>125 | 126-<br>138.1 | 138.2-<br>179 | 180-<br>223 | 224-<br>266 | 267-<br>340 | 341-<br>1397 | Row<br>Total |
|---|---|---|---|---|---|---|---|---|---|---|---|
| Remainder | 16<br>94.1 | 9<br>64.3 | 11<br>61.1 | 8<br>53.3 | 6<br>42.9 | 4<br>25.0 | 5<br>29.4 | 4<br>25.0 | 1<br>6.7 | 1<br>5.9 | 65<br>40.9 |
| Stage 2 | 1<br>5.9 | 2<br>14.3 | 6<br>33.3 | 5<br>33.3 | 4<br>28.6 | 4<br>25.0 | 7<br>41.2 | 3<br>18.8 | 5<br>33.3 | 4<br>23.5 | 41<br>25.8 |
| Stage 1 | 0<br>0.0 | 3<br>21.4 | 1<br>5.6 | 2<br>13.3 | 4<br>28.6 | 8<br>50.0 | 5<br>29.4 | 9<br>56.3 | 9<br>60.0 | 12<br>70.6 | 53<br>33.3 |
| Column<br>Total | 17<br>10.7 | 14<br>8.8 | 18<br>11.3 | 15<br>9.4 | 14<br>8.8 | 16<br>10.1 | 17<br>10.7 | 16<br>10.1 | 15<br>9.4 | 17<br>10.7 | 159<br>100.0 |

TABLE A.1. *Continued*

*12) U.S. 159*

Net Farm Income Q

| Count<br>Col % | 0-<br>3000 | 3001-<br>6500 | 6501-<br>9500 | 9501-<br>44000 | Row<br>Total |
|---|---|---|---|---|---|
| Remainder | 23<br>60.5 | 16<br>39.0 | 12<br>35.3 | 10<br>24.4 | 61<br>39.6 |
| Stage 2 | 8<br>21.1 | 11<br>26.8 | 14<br>41.2 | 8<br>19.5 | 41<br>26.6 |
| Stage 1 | 7<br>18.4 | 14<br>34.1 | 8<br>23.5 | 23<br>56.1 | 52<br>33.8 |
| Column<br>Total | 38<br>24.7 | 41<br>26.6 | 34<br>22.1 | 41<br>26.6 | 154<br>100.0 |

Number of Missing Observations = 5

Net Farm Income T

| Count<br>Col % | 0-<br>3500 | 3501-<br>7500 | 7501-<br>10000 | 10001<br>44000 | Row<br>Total |
|---|---|---|---|---|---|
| Remainder | 24<br>53.3 | 19<br>44.2 | 12<br>33.3 | 6<br>20.0 | 61<br>39.6 |
| Stage 2 | 11<br>24.4 | 11<br>25.6 | 13<br>36.1 | 6<br>20.0 | 41<br>26.6 |
| Stage 1 | 10<br>22.2 | 13<br>30.2 | 11<br>30.6 | 18<br>60.0 | 52<br>33.8 |
| Column<br>Total | 45<br>29.2 | 43<br>27.9 | 36<br>23.4 | 30<br>19.5 | 154<br>100.0 |

Number of Missing Observations = 5

Net Farm Income D

| Count<br>Col % | 0-<br>1200 | 1201-<br>3000 | 3001-<br>4000 | 4001-<br>5500 | 5501-<br>7000 | 7001-<br>8500 | 8501-<br>10000 | 10001-<br>15000 | 15001-<br>44000 | Row<br>Total |
|---|---|---|---|---|---|---|---|---|---|---|
| Remainder | 12<br>75.0 | 11<br>50.0 | 8<br>47.1 | 4<br>30.8 | 7<br>38.9 | 6<br>33.3 | 7<br>35.0 | 3<br>18.8 | 3<br>21.4 | 61<br>39.6 |
| Stage 2 | 2<br>12.5 | 6<br>27.3 | 4<br>23.5 | 5<br>38.5 | 4<br>22.2 | 9<br>50.0 | 5<br>25.0 | 3<br>18.8 | 3<br>21.4 | 41<br>26.6 |
| Stage 1 | 2<br>12.5 | 5<br>22.7 | 5<br>29.4 | 4<br>30.8 | 7<br>38.9 | 3<br>16.7 | 8<br>40.0 | 10<br>62.5 | 8<br>57.1 | 52<br>33.8 |
| Column<br>Total | 16<br>10.4 | 22<br>14.3 | 17<br>11.0 | 13<br>8.4 | 18<br>11.7 | 18<br>11.7 | 20<br>13.0 | 16<br>10.4 | 14<br>9.1 | 154<br>100.0 |

Number of Missing Observations = 5

TABLE A.1. *Continued*

*13) U.S. 167*

Area Q

| Count<br>Col % | 0-<br>48 | 49-<br>83 | 84-<br>166 | 167-<br>891 | Row<br>Total |
|---|---|---|---|---|---|
| Remainder | 30<br>75.0 | 19<br>48.7 | 27<br>65.9 | 12<br>31.6 | 88<br>55.7 |
| Stage 2 | 5<br>12.5 | 15<br>38.5 | 7<br>17.1 | 12<br>31.6 | 39<br>24.7 |
| Stage 1 | 5<br>12.5 | 5<br>12.8 | 7<br>17.1 | 14<br>36.8 | 31<br>19.6 |
| Column<br>Total | 40<br>25.3 | 39<br>24.7 | 41<br>25.9 | 38<br>24.1 | 158<br>100.0 |

Number of Missing Observations = 9

Area T

| Count<br>Col % | 0-<br>58 | 59-<br>110.2 | 110.6-<br>224 | 225-<br>891 | Row<br>Total |
|---|---|---|---|---|---|
| Remainder | 34<br>72.3 | 28<br>56.0 | 16<br>50.0 | 10<br>34.5 | 88<br>55.7 |
| Stage 2 | 6<br>12.8 | 17<br>34.0 | 6<br>18.8 | 10<br>34.5 | 39<br>24.7 |
| Stage 1 | 7<br>14.9 | 5<br>10.0 | 10<br>31.3 | 9<br>31.0 | 31<br>19.6 |
| Column<br>Total | 47<br>29.7 | 50<br>31.6 | 32<br>20.3 | 29<br>18.4 | 158<br>100.0 |

Number of Missing Observations = 9

Area D

| Count<br>Col % | 0-<br>32 | 33-<br>44 | 45-<br>57 | 58-<br>67 | 68-<br>83 | 84-<br>110.2 | 110.3-<br>159 | 160-<br>224 | 225-<br>267 | 268-<br>891 | Row<br>Total |
|---|---|---|---|---|---|---|---|---|---|---|---|
| Remainder | 12<br>66.7 | 12<br>85.7 | 10<br>66.7 | 7<br>41.2 | 8<br>53.3 | 13<br>72.2 | 8<br>57.1 | 8<br>44.4 | 5<br>31.3 | 5<br>38.5 | 88<br>55.7 |
| Stage 2 | 2<br>11.1 | 1<br>7.1 | 3<br>20.0 | 9<br>52.9 | 5<br>33.3 | 3<br>16.7 | 3<br>21.4 | 3<br>16.7 | 7<br>43.8 | 3<br>23.1 | 39<br>24.7 |
| Stage 1 | 4<br>22.2 | 1<br>7.1 | 2<br>13.3 | 1<br>5.9 | 2<br>13.3 | 2<br>11.1 | 3<br>21.4 | 7<br>38.9 | 4<br>25.0 | 5<br>38.5 | 31<br>19.6 |
| Column<br>Total | 18<br>11.4 | 14<br>8.9 | 15<br>9.5 | 17<br>10.8 | 15<br>9.5 | 18<br>11.4 | 14<br>8.9 | 18<br>11.4 | 16<br>10.1 | 13<br>8.2 | 158<br>100.0 |

Number of Missing Observations = 9

**TABLE A.1.** *Continued*

*13) U.S. 167*

Net Farm Income Q

| Count<br>Col % | 0-<br>500 | 501-<br>1300 | 1301-<br>2500 | 2501-<br>14500 | Row<br>Total |
|---|---|---|---|---|---|
| Remainder | 20<br>57.1 | 22<br>68.8 | 29<br>60.4 | 13<br>38.2 | 84<br>56.4 |
| Stage 2 | 7<br>20.0 | 4<br>12.5 | 12<br>25.0 | 15<br>44.1 | 38<br>25.5 |
| Stage 1 | 8<br>22.9 | 6<br>18.8 | 7<br>14.6 | 6<br>17.6 | 27<br>18.1 |
| Column<br>Total | 35<br>23.5 | 32<br>21.5 | 48<br>32.2 | 34<br>22.8 | 149<br>100.0 |

Number of Missing Observations = 18

Net Farm Income T

| Count<br>Col % | 0-<br>900 | 901-<br>1500 | 1501-<br>3100 | 3101-<br>14500 | Row<br>Total |
|---|---|---|---|---|---|
| Remainder | 27<br>60.0 | 35<br>70.0 | 13<br>46.4 | 9<br>34.6 | 84<br>56.4 |
| Stage 2 | 8<br>17.8 | 8<br>16.0 | 11<br>39.3 | 11<br>42.3 | 38<br>25.5 |
| Stage 1 | 10<br>22.2 | 7<br>14.0 | 4<br>14.3 | 6<br>23.1 | 27<br>18.1 |
| Column<br>Total | 45<br>30.2 | 50<br>33.6 | 28<br>18.8 | 26<br>17.4 | 149<br>100.0 |

Number of Missing Observations = 18

Net Farm Income D

| Count<br>Col % | 0 only | 75-<br>500 | 501-<br>1000 | 1001-<br>1500 | 1501-<br>2500 | 2501-<br>3500 | 3501-<br>14500 | Row<br>Total |
|---|---|---|---|---|---|---|---|---|
| Remainder | 10<br>55.6 | 10<br>58.8 | 19<br>70.4 | 23<br>69.7 | 9<br>45.0 | 5<br>33.3 | 8<br>42.1 | 84<br>56.4 |
| Stage 2 | 4<br>22.2 | 3<br>17.6 | 3<br>11.1 | 6<br>18.2 | 7<br>35.0 | 8<br>53.3 | 7<br>36.8 | 38<br>25.5 |
| Stage 1 | 4<br>22.2 | 4<br>23.5 | 5<br>18.5 | 4<br>12.1 | 4<br>20.0 | 2<br>13.3 | 4<br>21.1 | 27<br>18.1 |
| Column<br>Total | 18<br>12.1 | 17<br>11.4 | 27<br>18.1 | 33<br>22.1 | 20<br>13.4 | 15<br>10.1 | 19<br>12.8 | 149<br>100.0 |

Number of Missing Observations = 18

TABLE A.1. *Continued*

*14) India 676*

Area Q

| Count<br>Col % | 0-<br>15.9 | 16.0-<br>27.9 | 28.0-<br>54.9 | 55<br>and over | Row<br>Total |
|---|---|---|---|---|---|
| Remainder | 91<br>57.2 | 99<br>54.7 | 85<br>51.2 | 66<br>38.8 | 341<br>50.4 |
| Stage 2 | 48<br>30.2 | 46<br>25.4 | 50<br>30.1 | 36<br>21.2 | 180<br>26.6 |
| Stage 1 | 20<br>12.6 | 36<br>19.9 | 31<br>18.7 | 68<br>40.0 | 155<br>22.9 |
| Column<br>Total | 159<br>23.5 | 181<br>26.8 | 166<br>24.6 | 170<br>25.1 | 676<br>100.0 |

Area T

| Count<br>Col % | 0-<br>16.3 | 16.4-<br>32.6 | 32.7-<br>64.9 | 65.0<br>and over | Row<br>Total |
|---|---|---|---|---|---|
| Remainder | 114<br>55.9 | 103<br>50.5 | 79<br>56.0 | 45<br>35.4 | 341<br>50.4 |
| Stage 2 | 60<br>29.4 | 60<br>29.4 | 34<br>24.1 | 26<br>20.5 | 180<br>26.6 |
| Stage 1 | 30<br>14.7 | 41<br>20.1 | 28<br>19.9 | 56<br>44.1 | 155<br>22.9 |
| Column<br>Total | 204<br>30.2 | 204<br>30.2 | 141<br>20.9 | 127<br>18.8 | 676<br>100.0 |

Area D

| Count<br>Col % | 0-<br>11.9 | 12.0-<br>12.4 | 12.5-<br>16.7 | 16.8-<br>21.1 | 21.2-<br>28.0 | 28.1-<br>33.3 | 33.4-<br>48.3 | 48.4-<br>68.5 | 68.6-<br>100.9 | 101.0<br>and over | Row<br>Total |
|---|---|---|---|---|---|---|---|---|---|---|---|
| Remainder | 35<br>61.4 | 38<br>55.5 | 41<br>51.9 | 42<br>56.0 | 34<br>55.7 | 27<br>39.7 | 36<br>54.5 | 43<br>57.3 | 29<br>47.5 | 16<br>24.2 | 341<br>50.4 |
| Stage 2 | 17<br>29.8 | 21<br>30.9 | 22<br>27.8 | 19<br>25.3 | 15<br>24.6 | 26<br>38.2 | 19<br>28.8 | 15<br>20.0 | 10<br>16.4 | 16<br>24.2 | 180<br>26.6 |
| Stage 1 | 5<br>8.8 | 9<br>13.2 | 16<br>20.3 | 14<br>18.7 | 12<br>19.7 | 15<br>22.1 | 11<br>16.7 | 17<br>22.7 | 22<br>36.1 | 34<br>51.5 | 155<br>22.9 |
| Column<br>Total | 57<br>8.4 | 68<br>10.1 | 79<br>11.7 | 75<br>11.1 | 61<br>9.0 | 68<br>10.1 | 66<br>9.8 | 75<br>11.1 | 61<br>9.0 | 66<br>9.8 | 676<br>100.0 |

Number of Missing Observations = 9

TABLE A.1. *Continued*

*14) India 676*

Value Product Q

| Count<br>Col % | 0-<br>941 | 942-<br>2276 | 2277-<br>4578 | 4578<br>and over | Row<br>Total |
|---|---|---|---|---|---|
| Remainder | 138<br>82.1 | 111<br>65.7 | 63<br>37.1 | 29<br>17.4 | 341<br>50.6 |
| Stage 2 | 27<br>16.1 | 39<br>23.1 | 66<br>38.8 | 48<br>28.7 | 180<br>26.7 |
| Stage 1 | 3<br>1.8 | 19<br>11.2 | 41<br>24.1 | 90<br>53.9 | 153<br>22.7 |
| Column<br>Total | 168<br>24.9 | 169<br>25.1 | 170<br>25.2 | 167<br>24.8 | 674<br>100.0 |

Value Product T

| Count<br>Col % | 0-<br>1127 | 1128-<br>2911 | 2912-<br>5560 | 5561<br>and over | Row<br>Total |
|---|---|---|---|---|---|
| Remainder | 166<br>81.8 | 112<br>55.7 | 44<br>32.6 | 19<br>14.1 | 341<br>50.6 |
| Stage 2 | 32<br>15.8 | 61<br>30.3 | 51<br>37.8 | 36<br>26.7 | 180<br>26.7 |
| Stage 1 | 5<br>2.5 | 28<br>13.9 | 40<br>29.6 | 80<br>59.3 | 153<br>22.7 |
| Column<br>Total | 203<br>30.1 | 201<br>29.8 | 135<br>20.0 | 135<br>20.0 | 674<br>100.0 |

Value Product D

| Count<br>Row %<br>Col %<br>Tot % | 0-<br>557 | 558-<br>839 | 840-<br>1126 | 1127-<br>1606 | 1607-<br>2244 | 2245-<br>2897 | 2898-<br>3903 | 3904-<br>5524 | 5525-<br>8830 | 8831<br>and over | Row<br>Total |
|---|---|---|---|---|---|---|---|---|---|---|---|
| Remainder | 53<br>79.1 | 57<br>85.1 | 54<br>80.6 | 44<br>65.7 | 40<br>59.7 | 28<br>41.8 | 26<br>38.2 | 20<br>29.9 | 17<br>25.4 | 2<br>2.9 | 341<br>50.6 |
| Stage 2 | 13<br>19.4 | 9<br>13.4 | 10<br>14.9 | 19<br>28.4 | 14<br>20.9 | 28<br>41.8 | 25<br>36.8 | 26<br>38.8 | 20<br>29.9 | 16<br>22.9 | 180<br>26.7 |
| Stage 1 | 1<br>1.5 | 1<br>1.5 | 3<br>4.5 | 4<br>6.0 | 13<br>19.4 | 11<br>16.4 | 17<br>25.0 | 21<br>31.3 | 30<br>44.8 | 52<br>74.3 | 153<br>22.7 |
| Column<br>Total | 67<br>9.9 | 67<br>9.9 | 67<br>9.9 | 67<br>9.9 | 67<br>9.9 | 67<br>9.9 | 68<br>10.1 | 67<br>9.9 | 67<br>9.9 | 70<br>10.4 | 674<br>100.0 |

TABLE A.1. *Continued*

*15) India 1145*

Area Q

| Count<br>Col % | .04-<br>1.6 | 1.7-<br>3.04 | 3.08-<br>5.5 | 5.7-<br>364.2 | Row<br>Total |
|---|---|---|---|---|---|
| Remainder | 214<br>66.5 | 134<br>53.4 | 112<br>39.0 | 75<br>26.4 | 535<br>46.8 |
| Stage 2 | 85<br>26.4 | 78<br>31.1 | 100<br>34.8 | 74<br>26.1 | 337<br>29.5 |
| Stage 1 | 23<br>7.1 | 39<br>15.5 | 75<br>26.1 | 135<br>47.5 | 272<br>23.8 |
| Column<br>Total | 322<br>28.1 | 251<br>21.9 | 287<br>25.1 | 284<br>24.8 | 1144<br>100.0 |

Number of Missing Observations = 1

Area T

| Count<br>Col % | .04-<br>1.82 | 1.86-<br>3.9 | 4.0-<br>6.5 | 6.6-<br>364.2 | Row<br>Total |
|---|---|---|---|---|---|
| Remainder | 227<br>65.6 | 169<br>49.7 | 84<br>36.1 | 55<br>24.4 | 535<br>46.8 |
| Stage 2 | 92<br>26.6 | 111<br>32.6 | 82<br>35.2 | 52<br>23.1 | 337<br>29.5 |
| Stage 1 | 27<br>7.8 | 60<br>17.6 | 67<br>28.8 | 118<br>52.4 | 272<br>23.8 |
| Column<br>Total | 346<br>30.2 | 340<br>29.7 | 233<br>20.4 | 225<br>19.7 | 1144<br>100.0 |

Number of Missing Observations = 1

Area D

| Count<br>Col % | .04-<br>.7 | .8-<br>1.3 | 1.4-<br>1.82 | 1.86-<br>2.3 | 2.4-<br>3.03 | 3.07-<br>3.9 | 4.0-<br>4.8 | 5.1-<br>6.4 | 6.6-<br>10.1 | 10.3-<br>364.2 | Row<br>Total |
|---|---|---|---|---|---|---|---|---|---|---|---|
| Remainder | 78<br>75.0 | 80<br>65.0 | 69<br>58.0 | 58<br>61.7 | 63<br>47.4 | 48<br>42.5 | 43<br>31.2 | 41<br>43.2 | 30<br>26.5 | 25<br>22.3 | 535<br>46.8 |
| Stage 2 | 25<br>24.0 | 32<br>26.0 | 35<br>29.4 | 25<br>26.6 | 46<br>34.6 | 40<br>35.4 | 51<br>37.0 | 31<br>32.6 | 29<br>25.7 | 23<br>20.5 | 337<br>29.5 |
| Stage 1 | 1<br>1.0 | 11<br>8.9 | 15<br>12.6 | 11<br>11.7 | 24<br>18.0 | 25<br>22.1 | 44<br>31.9 | 23<br>24.2 | 54<br>47.8 | 64<br>57.1 | 272<br>23.8 |
| Column | 104<br>9.1 | 123<br>10.8 | 119<br>10.4 | 94<br>8.2 | 133<br>11.6 | 113<br>9.9 | 138<br>12.1 | 95<br>8.3 | 113<br>9.9 | 112<br>9.8 | 1144<br>100.0 |

Number of Missing Observations = 1

TABLE A.1. *Continued*

*15) India 1145*

Product Q

| Count<br>Col % | 30-<br>1348 | 1350-<br>3000 | 3008-<br>6360 | 6390-<br>86220 | Row<br>Total |
|---|---|---|---|---|---|
| Remainder | 226<br>78.2 | 150<br>53.0 | 109<br>38.1 | 50<br>17.5 | 535<br>46.8 |
| Stage 2 | 55<br>19.0 | 86<br>30.4 | 109<br>38.1 | 87<br>30.4 | 337<br>29.5 |
| Stage 1 | 8<br>2.8 | 47<br>16.6 | 68<br>23.8 | 149<br>52.1 | 272<br>23.8 |
| Column<br>Total | 289<br>25.3 | 283<br>24.7 | 286<br>25.0 | 286<br>25.0 | 1144<br>100.0 |

Number of Missing Observations = 1

Value Product T

| Count<br>Col % | 30-<br>1559 | 1560-<br>4070 | 4080-<br>8000 | 8050-<br>86220 | Row<br>Total |
|---|---|---|---|---|---|
| Remainder | 258<br>75.2 | 170<br>49.6 | 71<br>31.0 | 36<br>15.7 | 535<br>46.8 |
| Stage 2 | 68<br>19.8 | 115<br>33.5 | 93<br>40.6 | 61<br>26.6 | 337<br>29.5 |
| Stage 1 | 17<br>5.0 | 58<br>16.9 | 65<br>28.4 | 132<br>57.6 | 272<br>23.8 |
| Column<br>Total | 343<br>30.0 | 343<br>30.0 | 229<br>20.0 | 229<br>20.0 | 1144<br>100.0 |

Number of Missing Observations = 1

Value Product D

| Count<br>Col % | 30-<br>638 | 640-<br>1080 | 1090-<br>1559 | 1560-<br>2251 | 2280-<br>3000 | 3008-<br>4070 | 4080-<br>5485 | 5500-<br>8000 | 8050-<br>12747 | 12905-<br>86220 | Row<br>Total |
|---|---|---|---|---|---|---|---|---|---|---|---|
| Remainder | 98<br>86.7 | 83<br>71.6 | 77<br>67.5 | 69<br>60.0 | 49<br>43.0 | 52<br>45.6 | 41<br>35.7 | 30<br>26.3 | 22<br>19.1 | 14<br>12.3 | 535<br>46.8 |
| Stage 2 | 14<br>12.4 | 28<br>24.1 | 26<br>22.8 | 30<br>26.1 | 43<br>37.7 | 42<br>36.8 | 48<br>41.7 | 45<br>39.5 | 35<br>30.4 | 26<br>22.8 | 337<br>29.5 |
| Stage 1 | 1<br>0.9 | 5<br>4.3 | 11<br>9.6 | 16<br>13.9 | 22<br>19.3 | 20<br>17.5 | 26<br>22.6 | 39<br>34.2 | 58<br>50.4 | 74<br>64.9 | 272<br>23.8 |
| Column<br>Total | 113<br>9.9 | 116<br>10.1 | 114<br>10.0 | 115<br>10.1 | 114<br>10.0 | 114<br>10.0 | 115<br>10.1 | 114<br>10.0 | 115<br>10.1 | 114<br>10.0 | 1144<br>100.0 |

Number of Missing Observations = 1

**TABLE A.1.** *Continued*

*16) U.S. 316*

Gross Farm Income Q

| Count<br>Col % | 1-<br>5499 | 5500-<br>9499 | 9500-<br>13499 | 13500-<br>and over | Row<br>Total |
|---|---|---|---|---|---|
| Remainder | 47<br>79.7 | 53<br>58.9 | 40<br>45.5 | 17<br>21.5 | 157<br>49.7 |
| Stage 2 | 5<br>8.5 | 24<br>26.7 | 26<br>29.5 | 23<br>29.1 | 78<br>24.7 |
| Stage 1 | 7<br>11.9 | 13<br>14.4 | 22<br>25.0 | 39<br>49.4 | 81<br>25.6 |
| Column | 59<br>18.7 | 90<br>28.5 | 88<br>27.8 | 79<br>25.0 | 316<br>100.0 |

Gross Farm Income T

| Count<br>Col % | 1-<br>7999 | 8000-<br>11999 | 12000-<br>15999 | 16000<br>and over | Row<br>Total |
|---|---|---|---|---|---|
| Remainder | 75<br>70.8 | 49<br>54.4 | 25<br>39.1 | 8<br>14.3 | 157<br>49.7 |
| Stage 2 | 17<br>16.0 | 25<br>27.8 | 18<br>28.1 | 18<br>32.1 | 78<br>24.7 |
| Stage 1 | 14<br>13.2 | 16<br>17.8 | 21<br>32.8 | 30<br>53.6 | 81<br>25.6 |
| Column<br>Total | 106<br>33.5 | 90<br>28.5 | 64<br>20.3 | 56<br>17.7 | 316<br>100.0 |

Gross Farm Income D

| Count<br>Col % | 1-<br>3999 | 4000-<br>5499 | 5500-<br>7999 | 8000-<br>9499 | 9500-<br>11999 | 12000-<br>13499 | 13500-<br>19999 | 20000<br>and over | Row<br>Total |
|---|---|---|---|---|---|---|---|---|---|
| Remainder | 11<br>73.3 | 36<br>81.8 | 28<br>59.6 | 25<br>58.1 | 24<br>51.1 | 16<br>39.0 | 14<br>28.6 | 3<br>10.0 | 157<br>49.7 |
| Stage 2 | 3<br>20.0 | 2<br>4.5 | 12<br>25.5 | 12<br>27.9 | 13<br>27.7 | 13<br>31.7 | 18<br>36.7 | 5<br>16.7 | 78<br>24.7 |
| Stage 1 | 1<br>6.7 | 6<br>13.6 | 7<br>14.9 | 6<br>14.0 | 10<br>21.3 | 12<br>29.3 | 17<br>34.7 | 22<br>73.3 | 81<br>25.6 |
| Column<br>Total | 15<br>4.7 | 44<br>13.9 | 47<br>14.9 | 43<br>13.6 | 47<br>14.9 | 41<br>13.0 | 49<br>15.5 | 30<br>9.5 | 316<br>100.0 |

Number of Missing Observations = 1

TABLE A.2. *Standardized Adoption Rates and Resulting Tests of Hypotheses*

| | CUT[a] | IV[b] | STAGE 1 L | STAGE 1 LM | STAGE 1 HM | STAGE 1 H | STAGE 2 L | STAGE 2 LM | STAGE 2 HM | STAGE 2 H | Hypothesis A[c] | Hypothesis B[c] |
|---|---|---|---|---|---|---|---|---|---|---|---|---|
| INDIA 185 | Q | A | 0 | 21 | 15 | 64 | 21 | 29 | 19 | 32 | X[d] | 0[c] |
| | T | A | 5 | 19 | 22 | 55 | 20 | 25 | 23 | 32 | 0 | 0 |
| INDIA 412 | Q | A | 8 | 29 | 31 | 33 | 23 | 27 | 21 | 30 | 0 | 0 |
| | T | A | 8 | 28 | 26 | 38 | 22 | 23 | 24 | 32 | X | X |
| | Q | I | 13 | 11 | 30 | 42 | 19 | 22 | 29 | 31 | 0 | 0 |
| | T | I | 12 | 14 | 31 | 43 | 18 | 25 | 31 | 26 | 0 | 0 |
| KENYA 540 | Q | I | 19 | 24 | 27 | 30 | 17 | 27 | 30 | 26 | 0 | 0[e] |
| | T | I | 18 | 27 | 25 | 30 | 17 | 25 | 30 | 28 | X | X |
| MEXICO 108 | Q | A | 16 | 22 | 5 | 58 | 0 | 18 | 44 | 38 | X | X |
| MEXICO 193 | Q | A | 12 | 27 | 24 | 38 | 20 | 24 | 22 | 35 | X | X |
| PAKISTAN 221 | Q | A | 18 | 23 | 25 | 35 | 22 | 27 | 28 | 22 | 0 | 0 |
| | T | A | 17 | 24 | 18 | 41 | 22 | 27 | 28 | 23 | X | X |
| PAKISTAN 350 | Q | A | 9 | 26 | 14 | 51 | 23 | 24 | 26 | 28 | X | X |
| | T | A | 9 | 24 | 15 | 51 | 23 | 23 | 25 | 28 | X | X |
| PHILIPPINES 153 | Q | A | 19 | 14 | 24 | 43 | 26 | 26 | 22 | 27 | 0 | 0 |
| | T | A | 17 | 21 | 18 | 44 | 28 | 27 | 18 | 27 | X | 0 |
| | Q | I | 10 | 25 | 27 | 38 | 28 | 29 | 13 | 29 | 0 | 0 |
| | T | I | 10 | 27 | 26 | 37 | 25 | 19 | 20 | 36 | X | X |
| PHILIPPINES 177 | Q | A | 20 | 27 | 26 | 27 | 23 | 20 | 27 | 30 | X | X |
| | T | A | 19 | 27 | 26 | 27 | 23 | 19 | 27 | 30 | X | X |
| | Q | I | 13 | 20 | 32 | 35 | 15 | 24 | 35 | 28 | 0 | 0 |
| | T | I | 12 | 29 | 25 | 34 | 15 | 29 | 27 | 29 | X | X |
| TAIWAN 159 | T | A | 21 | 29 | 27 | 23 | 14 | 21 | 36 | 30 | X | X |
| TAIWAN 237 | Q | A | 19 | 31 | 26 | 23 | 15 | 26 | 23 | 37 | X | X[e] |
| U.S. 159 | Q | A | 8 | 13 | 32 | 47 | 8 | 22 | 32 | 38 | 0 | 0 |
| | T | A | 6 | 21 | 29 | 45 | 15 | 32 | 40 | 14 | 0 | 0[e] |
| | Q | I | 14 | 26 | 18 | 42 | 17 | 19 | 35 | 29 | X | X |
| | T | I | 11 | 21 | 21 | 42 | 18 | 22 | 31 | 29 | 0 | X |

TABLE A.2. *Continued*

| | CUT[a] | IV[b] | STAGE 1 L | LM | HM | H | STAGE 2 L | LM | HM | H | Hypothesis A[a] | Hypothesis B[a] |
|---|---|---|---|---|---|---|---|---|---|---|---|---|
| U.S. 167 | Q | A | 16 | 16 | 22 | 46 | 11 | 34 | 16 | 39 | 0 | 0 |
| | T | A | 17 | 11 | 36 | 36 | 12 | 29 | 21 | 38 | 0 | 0 |
| | Q | I | 31 | 25 | 20 | 24 | 21 | 12 | 24 | 43 | X | X |
| | T | I | 30 | 19 | 19 | 31 | 16 | 15 | 32 | 39 | 0 | X |
| INDIA 676 | Q | A | 14 | 22 | 21 | 44 | 25 | 23 | 27 | 25 | X | X |
| | T | A | 15 | 20 | 20 | 45 | 25 | 27 | 22 | 27 | X | 0 |
| | Q | I | 2 | 12 | 26 | 59 | 11 | 17 | 33 | 40 | 0 | X |
| | T | I | 2 | 13 | 28 | 56 | 9 | 21 | 37 | 38 | 0 | 0 |
| INDIA 1145 | Q | A | 7 | 16 | 27 | 49 | 18 | 23 | 29 | 31 | 0 | 0 |
| | T | A | 7 | 17 | 27 | 49 | 17 | 24 | 30 | 29 | 0 | 0 |
| | Q | I | 3 | 17 | 25 | 55 | 12 | 21 | 29 | 37 | 0 | X |
| | T | I | 5 | 16 | 26 | 53 | 12 | 22 | 31 | 35 | 0 | 0 |
| U.S. 316 | Q | I | 12 | 14 | 25 | 49 | 7 | 23 | 29 | 42 | 0 | 0 |
| | T | I | 11 | 15 | 28 | 46 | 11 | 21 | 26 | 42 | 0 | 0 |
| MEXICO 93 | Q | A | 17 | 21 | 16 | 46 | 0 | 15 | 36 | 49 | X | X |
| JAPAN 91 | Q | A | 18 | 27 | 23 | 33 | 20 | 28 | 29 | 22 | X | X |
| U.S. 173 | Q | I | 16 | 22 | 19 | 43 | 8 | 17 | 37 | 39 | X | X |
| U.S. 251 | Q | I | 13 | 23 | 27 | 36 | 24 | 27 | 26 | 23 | 0 | 0 |
| U.S. 252 | Q | A | 23 | 13 | 24 | 40 | 27 | 11 | 26 | 36 | 0 | X |
| U.S. 341 | Q | A | 10 | 11 | 25 | 54 | 17 | 18 | 23 | 42 | 0 | 0 |
| U.S. 413 | Q | A | 14 | 30 | 26 | 30 | 10 | 27 | 32 | 32 | X | X |

a. Cutting points on independent variable are either Quartile (Q) or "30/20" (T).

b. The independent variable is either A (land area) or I (income or value of product). See Table 4.1 for details.

c. Where the rounded numbers shown indicate a tie, these column indicate the outcome using unrounded calculations.

d. X = confirms hypothesis; 0 = does not confirm hypothesis.

e. In these cases (Kenya 540, 25/25; Taiwan 237; and U.S. 159, area, 30/20) the result is different using standardized (three confirmations) and raw (no confirmations) adoption rates. See the text for a discussion of alternative specifications. As is explained there, for the purposes of tabulating results, Taiwan is counted as confirming Hypothesis B, the other two cases as not confirming.

## *Appendix B*

# *The Use of Continuous Variables: Issues, Illustrations, and Scattergrams*

There are at least three reasons why an analysis that uses continuous measurement of variables may be preferable to the crosstabular techniques used in the body of this book.

1) The theory stated in Chapter 2 does not specify exactly where inflections should appear in the curvilinear relationships it predicts. Thus, while the cutting points used in the crosstabular analysis are appropriate specifications of the theory, in some cases they may lead to averaging over actual inflections that would be picked up by other appropriate specifications of the theory. Given the diversity of settings from which the data are taken and the uniformity of the analysis, it is likely that such inflections are obscured in at least some of the data sets reviewed. The proper continuous measurement would reveal the actual pattern in the data.

2) The use of crosstabular analysis (ordinal measurement) when interval or ratio scale measurement of the variables is available is commonly thought to throw away information, and consequently to confine the analyst to less revealing and powerful analysis. Continuous variables are usually associated with the more powerful statistical techniques.

3) The relationship between rank and innovation described in Figure 1 resembles the prototypical shape described by a cubic (third-degree) polynomial regression equation ($y = a + b_1x + b_2x^2 + b_3x^3$, see Blalock 1972: 460). Since this shape is one of the most unusual implications of the theory (the other being the difference between stages), fitting the distribution of innovation over rank in Stage 1 with a cubic regression equation would lend both elegance and power to the analysis.

In their criticism and reassessment of my earlier publications, Gartrell, Wilkening, and Presser (1973) suggested and illustrated the use of least-squares regression in the study of curvilinear relationships.

These considerations compelled me to extensively explore various sorts of "continuous" analysis, including straightforward linear and polynomial regression, a percentile scale for the independent variable, a moving average rate for the dependent variable, and various combinations of them.

My conclusion is that, for me with these data at this time, regression analysis cannot clarify any major substantive points. This conclusion has led me to present scattergrams for a number of the cases in the last section of this appendix. They show the predicted curvilinearity in the Stage 1 relation of rank and adoption. What follows discusses the issues involved in formulation of appropriate continuous measures, explains and illustrates the exploration of regression analysis, and describes the selection and construction of the scattergrams.

### *Issues: The Scale of Variables*

Two issues that are central to the theory being tested in this study heavily influenced the approach taken to continuous measures of the variables. The first issue involves the scale of the independent variable. In Chapter 4 I suggested that relative measures involving people are often more appropriate stratification variables than absolute measures involving underlying variables like area operated. In the present context an extension of that argument suggests that a person, rather than a hectare, peso, dollar or rupee, be used as the unit of measurement. Following this suggestion means that the percentile rank of an individual provides an interval rather than an ordinal measurement scale. In fact, if community of reference is properly defined, the percentile rank provides a ratio scale. The reconceptualization makes the difference between the "underlying variable" and the "rank variable" a substantive rather than a statistical or technical one.

In what follows I will maintain the distinction between "raw" and "percentiled" independent variables. The "raw" variable uses the underlying scale of money, land area, and so on, as the metric for calculation. The "percentiled" variable, while maintaining the individuals in the same rank order based on money or land, uses the percentile rank within the sample as the metric for calculation. That is, the percentile variable is not based on how much income the individual has, but on how many people in the community have less (or more) income than he or she does. While transformation of the raw variable into its percentiled version changes no one's rank, it spreads the cases differently in a bivariate plot. This, as will become apparent, can present a different picture for regression analysis or the eye.

The second issue is the scale of the dependent variable. The simplest specification of the theory leads to a two-value (dummy) dependent variable for testing Hypothesis A. Either an individual is a Stage 1 adopter (among the first 25 percent to adopt) or he or she is not. However, the dummy variable does not represent the more sensible and realistic interpretation of the theory in which the curve in Figure 1 specifies how the probability that any individual will be in Stage 1 varies with the individual's rank. In operational terms, this probabilistic interpretation means that different rates of early (Stage 1) adoption should be found in different sections of the rank continuum, and these rates should vary continuously. Thus, while a dummy variable provides the simplest operationalization of the theory, a continuous variable might provide a more accurate one.

Membership in Stage 1 may be transformed into an approximately continuous variable in a variety of ways that involve taking a rate of Stage 1 membership among individuals in arbitrarily defined sections of the independent variable. Two such transformations are used below. The first, called "fixed five," divides the independent variable into about twenty parts, each including 5 percent of the farmers; and the rate of Stage 1 membership is calculated for each group of farmers. For example, in a community of 500 farmers, the poorest 25 (5 percent) would be at the lowest of 20 points on the independent variable. If 2 of the 25 adopted during Stage 1, the adoption rate would be 8 percent for Stage 1. This transformation is simply a finer division of the rank continuum than the decile division shown in Table A.1.

The second continuous dependent variable is called "moving average twenty." Stage 1 membership rates are calculated for groups of farmers defined by a "window" that encompasses 20 percent of the farmers and moves along the rank continuum in 1-percent steps. That is, the first calculation gives the proportion of farmers among the poorest 20 percent who are Stage 1 members. The second calculation leaves out the lowest 1 percent and includes the equivalent number from the rank order just above the highest included in the first calculation; and the eighty-first and final calculation is for the top 20 percent of the ordered cases. Each one of the eighty-one calculations provides a value on the dependent variable that can be used in scattergrams and regression analysis.

While the "fixed five" division provides a fairly conventional measure of the dependent variable (it merely lumps data), the "moving average twenty" measure, of course, stabilizes and smooths the data (since, on average, 95 percent of the data used to calculate one data point is also used for calculating adjacent data points).*

In sum, I have defined two representations of an independent variable: a "raw" one that uses the scale of the underlying variables as a metric, and a "percentiled" one that uses the percentile rank of the individual as the metric. There is a substantive difference between them; and the percentile variable is the more appropriate, given the theory being tested here. In parallel fashion I have defined three measures of the dependent variable: a dummy variable that records presence or absence of Stage 1 membership for each individual; a "fixed

*Both the fixed five and the moving average twenty intervals are applied after the independent variable has been transformed into a percentile scale that uses mean percentile rank to represent ties (Sailer 1977).

five" variable that lumps parts of the population and records a rate of Stage 1 memberships for the lump; and a "moving average twenty" variable that does a similar thing for overlapping 20-percent slices of the population. These three dependent variables are different approximations of the same concept of adoption rate; and there is no substantive difference between them.

The following section briefly illustrates the implications for regression analysis of using each of the definitions of the variables. Before proceeding I should note that this appendix does not explore two important issues that are related to the ones it does explore. First, the illustrations below consistently assume the 25 percent definition of Stage 1, even though there is no clear theoretical reason to think that it is better than, say, 15 percent. Second, implications of the theory summarized in Hypothesis B (that is, the relation of rank and adoption in Stage 2) are not considered here.

### *Illustrations*

These illustrations are meant mostly for the reader who is interested in the application of polynomial regression to problems of the kind addressed in this book. While they are principally intended to record a small part of the process by which I decided that polynomial regression analysis is too complicated and too little understood for a comparative study with substantive rather than methodological goals, they also show the importance of employing the "percentiled" as contrasted with the "raw" independent variable in any analysis of curvilinearity. The reader who is not interested in the material on polynomial regression analysis should study Figures B.2–B.4 and skip to the final section of this appendix.

To illustrate the various options for analysis I have constructed a fake data set (Ideal 100). With such a simulated data set the irrelevant idiosyncracies of natural data sets can be avoided, and characteristics of the analytic techniques can be seen and compared more clearly. To insure relevance to the natural world, one of the data sets (Pakistan 105) from among those used in the body of this book is included in some of the later illustrations.

The actual data in Ideal 100 is displayed in Table B.1. Ideal 100 has 100 cases for 2 independent variables: Percentile (PER) and Value (VAL). As the table shows, PER ranges from 1 to 100. It has 1 case for each digit in the range; its mean is 50.5; and it is equivalent to a rank order of the cases. VAL ranges from 2 to 99, has a mean of 19.7, and is similar to the positively skewed distributions of land and income described in Tables 6 and 7. It is more skewed than some and less skewed than others. For now, we may take PER to represent the ideal transformation of a "raw" variable into a percentiled variable and VAL to represent the raw variable. Later the raw VAL distribution will be percentiled to show the effect of that transformation on a variable with

TABLE B.1. *The Ideal 100 Data Set*

| PER | VAL | ADOP | REAL | OLD | HIPT | PER | VAL | ADOP | REAL | OLD | HIPT | PER | VAL | ADOP | REAL | OLD | HIPT | PER | VAL | ADOP | REAL | OLD | HIPT |
|---|---|---|---|---|---|---|---|---|---|---|---|---|---|---|---|---|---|---|---|---|---|---|---|
| 1 | 2 | 0 | 0 | 0 | 0 | 26 | 5 | 0 | 0 | 0 | 0 | 51 | 10 | 0 | 0 | 0 | 0 | 76 | 28 | 0 | 1 | 0 | 1 |
| 2 | 2 | 0 | 0 | 0 | 0 | 27 | 5 | 1 | 0 | 0 | 0 | 52 | 10 | 1 | 0 | 0 | 0 | 77 | 29 | 1 | 0 | 0 | 1 |
| 3 | 2 | 0 | 0 | 0 | 0 | 28 | 5 | 0 | 1 | 0 | 0 | 53 | 11 | 0 | 0 | 0 | 0 | 78 | 30 | 0 | 1 | 1 | 1 |
| 4 | 2 | 0 | 0 | 0 | 0 | 29 | 5 | 0 | 0 | 0 | 0 | 54 | 11 | 0 | 0 | 0 | 0 | 79 | 31 | 0 | 0 | 0 | 1 |
| 5 | 2 | 0 | 0 | 0 | 0 | 30 | 5 | 1 | 1 | 1 | 0 | 55 | 12 | 1 | 0 | 1 | 0 | 80 | 33 | 1 | 1 | 0 | 1 |
| 6 | 2 | 0 | 0 | 1 | 0 | 31 | 6 | 0 | 0 | 0 | 0 | 56 | 12 | 0 | 0 | 0 | 0 | 81 | 35 | 0 | 0 | 1 | 1 |
| 7 | 2 | 0 | 0 | 0 | 0 | 32 | 6 | 0 | 1 | 0 | 0 | 57 | 13 | 1 | 0 | 0 | 0 | 82 | 37 | 1 | 1 | 0 | 1 |
| 8 | 2 | 0 | 0 | 0 | 0 | 33 | 6 | 0 | 0 | 0 | 0 | 58 | 13 | 0 | 0 | 0 | 0 | 83 | 39 | 0 | 0 | 0 | 1 |
| 9 | 3 | 0 | 0 | 0 | 0 | 34 | 6 | 0 | 1 | 0 | 0 | 59 | 14 | 0 | 0 | 1 | 0 | 84 | 41 | 0 | 1 | 1 | 1 |
| 10 | 3 | 1 | 0 | 0 | 0 | 35 | 6 | 0 | 0 | 1 | 0 | 60 | 14 | 1 | 0 | 0 | 0 | 85 | 42 | 1 | 0 | 0 | 1 |
| 11 | 3 | 0 | 0 | 0 | 0 | 36 | 6 | 0 | 1 | 0 | 0 | 61 | 15 | 0 | 0 | 0 | 0 | 86 | 43 | 0 | 1 | 0 | 1 |
| 12 | 3 | 0 | 0 | 1 | 0 | 37 | 6 | 1 | 0 | 0 | 0 | 62 | 15 | 0 | 0 | 0 | 0 | 87 | 45 | 1 | 0 | 1 | 1 |
| 13 | 3 | 0 | 0 | 0 | 0 | 38 | 6 | 0 | 1 | 0 | 0 | 63 | 16 | 0 | 0 | 1 | 0 | 88 | 47 | 0 | 1 | 0 | 1 |
| 14 | 3 | 0 | 0 | 0 | 0 | 39 | 7 | 0 | 0 | 0 | 0 | 64 | 16 | 0 | 0 | 0 | 0 | 89 | 49 | 0 | 0 | 0 | 1 |
| 15 | 3 | 0 | 0 | 0 | 0 | 40 | 7 | 1 | 1 | 1 | 0 | 65 | 17 | 0 | 0 | 0 | 0 | 90 | 51 | 1 | 1 | 1 | 1 |
| 16 | 3 | 0 | 0 | 0 | 0 | 41 | 7 | 0 | 0 | 0 | 0 | 66 | 18 | 0 | 0 | 0 | 0 | 91 | 53 | 0 | 0 | 0 | 1 |
| 17 | 3 | 0 | 0 | 0 | 0 | 42 | 7 | 0 | 1 | 0 | 0 | 67 | 19 | 1 | 0 | 1 | 0 | 92 | 55 | 1 | 1 | 1 | 1 |
| 18 | 3 | 0 | 0 | 1 | 0 | 43 | 8 | 0 | 0 | 0 | 0 | 68 | 20 | 0 | 0 | 0 | 0 | 93 | 57 | 0 | 0 | 0 | 1 |
| 19 | 4 | 0 | 0 | 0 | 0 | 44 | 8 | 0 | 1 | 0 | 0 | 69 | 21 | 0 | 0 | 0 | 0 | 94 | 59 | 0 | 1 | 1 | 1 |
| 20 | 4 | 1 | 0 | 0 | 0 | 45 | 9 | 1 | 0 | 1 | 0 | 70 | 22 | 1 | 0 | 0 | 0 | 95 | 62 | 1 | 0 | 0 | 1 |
| 21 | 4 | 0 | 0 | 0 | 0 | 46 | 9 | 0 | 1 | 0 | 0 | 71 | 23 | 0 | 0 | 1 | 0 | 96 | 69 | 0 | 1 | 1 | 1 |
| 22 | 5 | 0 | 0 | 0 | 0 | 47 | 9 | 1 | 0 | 0 | 0 | 72 | 24 | 0 | 0 | 0 | 0 | 97 | 77 | 1 | 0 | 0 | 1 |
| 23 | 5 | 0 | 0 | 0 | 0 | 48 | 9 | 0 | 1 | 0 | 0 | 73 | 25 | 0 | 0 | 0 | 0 | 98 | 89 | 0 | 1 | 1 | 1 |
| 24 | 5 | 0 | 0 | 1 | 0 | 49 | 9 | 0 | 0 | 0 | 0 | 74 | 26 | 0 | 0 | 0 | 0 | 99 | 95 | 0 | 0 | 0 | 1 |
| 25 | 5 | 0 | 0 | 0 | 0 | 50 | 10 | 1 | 1 | 1 | 0 | 75 | 27 | 0 | 0 | 1 | 0 | 100 | 99 | 1 | 1 | 1 | 1 |

an initially skewed distribution; and at that point PER will serve only as a stable comparison point for the change in results effected in the analysis using VAL.

Ideal 100 is designed to illustrate the effects of various transformations of the independent variable on both "linear" and "curvilinear" distributions of the dependent variable. To that end it includes 4 distributions of the 25 adopters (Stage 1) and 75 nonadopters over the independent variable. The first distribution is called Adoption (ADOP) and represents the pattern predicted by the theory for Stage 1. The second distribution, called Real (REAL), is simply a clear, but unrealistic, exaggeration of the same pattern. The third distribution, Old (OLD), approximates the linear positive relation of adoption rate and wealth interpreted with the conventional wisdom. The fourth distribution, High Point (HIPT), exaggerates this Old pattern. A few minutes spent with Table B.1 is the best way to understand these distributions.

*Illustrations with dummy dependent variable.* Table B.2 shows the $R^2$ and the t ratios for six different regression equations for each combination of the two independent variables and the four dependent variables from Ideal 100. Figure B.1 shows the regression equations and plots for selected polynomial regressions. An intuitive grasp of the original distributions shown in Table B.1 is important for interpreting these results.

The results in the table and figure suggest that:

1) Even though the patterns of the distribution of ADOP, OLD, and especially REAL are clear, $R^2$ is lower than might be expected with a continuous dependent variable.

2) As is expectable, polynomial regression has more success with the curvilinear distributions (ADOP and REAL) than it does with the realistic linear one (OLD), and the opposite is true for the linear (single-term) regression.

3) The evenly distributed PER variable yields consistently higher $R^2$ than the positively skewed VAL variable with the third and fourth degree equations and the curvilinear dependent variables (ADOP and REAL).

4) The plot of a polynomial regression equation does not always resemble the original distribution in any simple sense.

These illustrations are provided to give the reader a sense for regression analysis of dummy dependent variables in the context of this study, which hypothesizes a curvilinear relationship between the principal variables. They highlight the need for the "fixed five" and "moving average twenty" transformations of the dependent variable discussed above and illustrated immediately below.

TABLE B.2. *Regression Analysis with Dummy Dependent Variable*

| | Ideal 100 PER | | Ideal 100 VAL | |
|---|---|---|---|---|
| | $R^2$ [a] | t Ratio[b] | $R^2$ | t Ratio[b] |
| ADOP | | | | |
| $x$ | .044 | 2.4 | .029 | 2.0 |
| $x+x^2$ | .035 | __ __ | .023 | 1.3 __ |
| $x+x^2+x^3$ | .032 | 1.1 __ __ | .016 | 1.0 __ __ |
| $x^2$ | .037 | 2.2 | .016 | 1.6 |
| $x^2+x^3$ | .029 | __ __ | .016 | 1.3 1.0 |
| $x^2+x^3+x^4$ | .034 | 1.4 1.3 1.2 | .011 | 1.1 __ __ |
| REAL | | | | |
| $x$ | .069 | 2.9 | .074 | 3.0 |
| $x+x^2$ | .059 | __ __ | .065 | 1.2 __ |
| $x+x^2+x^3$ | .114 | 2.7 2.6 2.6 | .057 | __ __ __ |
| $x^2$ | .065 | 2.8 | .060 | 2.7 |
| $x^2+x^3$ | .058 | __ __ | .066 | 1.9 1.3 |
| $x^2+x^3+x^4$ | .105 | 2.4 2.4 2.5 | .058 | __ __ __ |
| OLD | | | | |
| $x$ | .043 | 2.3 | .056 | 2.6 |
| $x+x^2$ | .041 | __ __ | .046 | 1.0 __ |
| $x+x^2+x^3$ | .037 | __ __ __ | .036 | __ __ __ |
| $x^2$ | .050 | 2.5 | .046 | 2.4 |
| $x^2+x^3$ | .043 | __ __ | .043 | 1.4 __ |
| $x^2+x^3+x^4$ | .036 | __ __ __ | .036 | __ __ __ |
| HIPT | | | | |
| $x$ | .558 | 11.2 | .718 | 15.9 |
| $x+x^2$ | .793 | 6.2 10.6 | .781 | 11.2 5.4 |
| $x+x^2+x^3$ | .796 | 1.0 __ 1.6 | .819 | __ 3.5 4.6 |
| $x^2$ | .714 | 15.0 | .504 | 10.1 |
| $x^2+x^3$ | .796 | 3.2 6.4 | .819 | 16.5 13.1 |
| $x^2+x^3+x^4$ | .814 | 4.0 4.0 3.2 | .843 | 9.2 5.7 4.0 |

[a] All the regression statistics in this report are calculated with the Jaguar regression program (Lave 1973). It adjusts $R^2$ for degrees of freedom.

[b] A __ indicates the t ratio is less than 1.

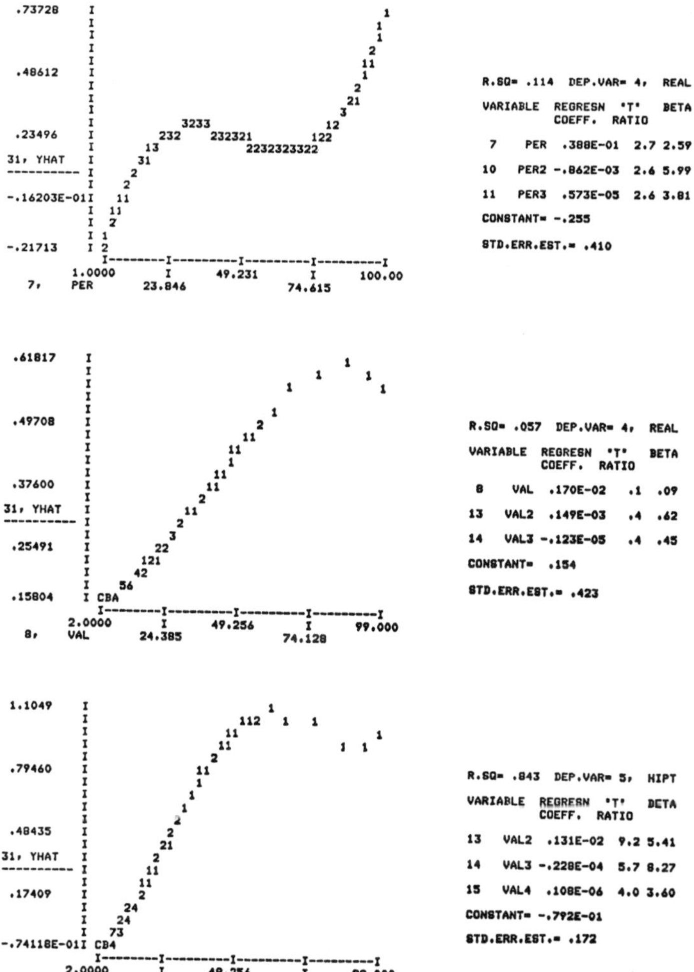


***Fig. B.1.** Plots of dummy variable regressions: Ideal 100, exaggerated distributions. The numbers indicate the number of "observations" at that point. Letters are used for numbers over nine: A = 10 to 15; B = 16 to 20; C = 21 to 25, etc.*

*Illustrations with continuous dependent variables.* The following illustrations are more directly relevant to the problem studied here. A continuous dependent variable is preferable to a dummy dependent variable for at least two reasons. First, the theory implies that the rate of innovation with respect to economic rank will increase and decrease in the continuous manner illustrated in Figure 1. Second, a continuous dependent variable permits the construction of scatter diagrams that can be judged as approximations to the hypothesized curve illustrated in Figure 1.*

The results of explorations using the two continuous dependent variables (fixed five and moving average twenty) and the two forms of the independent variable (raw and percentiled) are illustrated below. Note that, in order to connect this illustration using a fake data set with the analysis of natural situations, one of the studies from the body of this report is added to the analysis. The case selected, Pakistan 105, is a subsample separated out from Pakistan 221 because it represents a better approximation of a community of reference (see Chapter 6 and Appendix C).†

It is useful to look first at the scattergrams resulting from these transformations of the dependent variable. Figures B.2–4 show scattergrams for Ideal 100 with the PER independent variable, Ideal 100 with the VAL independent variable, and Pakistan 105. ADOP is the underlying dependent variable in the cases from Ideal 100. Each figure includes four scattergrams. For each measure of the dependent variable (fixed five and moving average twenty) there is a scattergram based on a percentiled independent variable and a scattergram based on a raw independent variable.

The raw independent variable needs explanation for the reader who

*I want to note in passing that desire for a continuous dependent variable led Gartrell, Wilkening, and Presser to misspecify the theory: "In an effort to measure innovation as a continuous variable we elected to use the sum of the total years since each innovation was first tried" (1973: 402). This measure merges Stage 1 and Stage 2 and thus cannot be considered a test of the hypothesis (A) about the relation of rank and innovation in Stage 1.

†Pakistan 105 is a good example because its size (N = 105) is close to that of Ideal 100 (N = 100). It is atypical of the natural data sets in that its scattergram yields an exceptionally clear confirmation of Hypothesis A.

In its "raw" form Pakistan 105 has 105 cases, a single independent variable (hectares of wheat harvested), and a single dependent variable (year of first use of HYV wheat seed). Here we will use two independent variables: the hectares measure (times 100) and the transformation of the hectares measure into a percentile rank variable. There are thirty-one farmers in Stage 1 and seventy-four farmers not in Stage 1.

A crosstabular description of Pakistan 105 that parallels the form of Table A.1 is found in Appendix C. It is worth contrasting the clear confirmation of Hypothesis A, shown here, with the ambiguity produced by the crosstabular presentation in Appendix C.

```
0.40000     I             1    1    1 1 1    1    1 1 1 1   1
            I
            I
            I
            I
0.29474     I
            I
            I
            I
            I
0.18947     I  1    1            1
            I
 3, ST1FX5  I
----------  I
            I
0.84211E-01 I
            I
            I
            I
0.00000E+00 I 1    1    1    1                 1    1
              I---------I---------I---------I---------I
         2.0000         I       48.282      I      97.00
   2, I100A%         23.923              72.641
```

Percentiled Fixed Five

```
0.40000     I                                              1221
            I
            I
            I
            I                 22321                   21
0.32105     I
            I
            I            12      1222        1222
            I
            I
0.24211     I        2221              21    21
            I
 3, ST1020  I
----------  I    2222               1222
            I
0.16316     I
            I 22222
            I
            I
0.10000E+00 I 322
              I---------I---------I---------I---------I
         9.5000         I       48.474      I      89.50
   2, I100A%         27.962              68.987
```

Percentiled Moving Average Twenty

```
0.40000     I             1    1    1 1 1    1    1 1 1 1   1
            I
            I
            I
            I
0.29474     I
            I
            I
            I
            I
0.18947     I  1    1            1
            I
 3, ST1FX5  I
----------  I
            I
0.84211E-01 I
            I
            I
            I
0.00000E+00 I 1    1    1    1                 1    1
              I---------I---------I---------I---------I
         3.0000         I       49.282      I      98.00
   1, I100A          24.923              73.641
```

Raw Fixed Five

```
0.40000     I                                              1221
            I
            I
            I
            I                 22321                   21
0.32105     I
            I
            I            12      1222        1222
            I
            I
0.24211     I        2221              21    21
            I
 3, ST1020  I
----------  I    2222               1222
            I
0.16316     I
            I 22222
            I
            I
0.10000E+00 I 322
              I---------I---------I---------I---------I
         10.500         I       49.474      I      90.50
   1, I100A          28.962              69.987
```

Raw Moving Average Twenty

***Fig. B.2.*** *Scattergrams: Ideal 100 per*

```
0.50000    I                       1
           I
           I
           I                     1
           I                  1        1   1 1 1  1  1
0.36842    I
           I
           I       1
           I
           I
0.23684    I                1
           I         2
 3, ST1FX5 I
---------- I
           I
0.10526    I             2
           I    2
           I
           I
0.00000E+00I 2                        1   1
             I---------I---------I---------I---------I
          3.5000       I       49.051      I       97.00
   2, I100I%       25.077              73.026

Percentiled Fixed Five
```

```
0.40000    I                                      1221
           I                    1
           I
           I                   1
           I                   2 22              21
0.31980    I                  2   1 11
           I                  1   1 1
           I                        1       1222
           I               2
           I                          11   1
0.23960    I       1        1         11   11
           I
 3, ST1020 I               4           11 1
---------- I         6                  111
           I     7     3
0.15940    I           3
           I  2                          1
           I              1
           I
0.95238E-01I 4
             I---------I---------I---------I---------I
           10.000      I       48.731      I       89.50
   2, I100I%       28.346              69.115

Percentiled Moving Average Twenty
```

```
0.50000    I      1
           I
           I
           I     1
           I    1    1    1   1  1    1              1
0.36842    I
           I
           I 1
           I
           I
0.23684    I   1
           I  2
 3, ST1FX5 I
---------- I
           I
0.10526    I  2
           I 2
           I
           I
0.00000E+00I 2     1   1
             I---------I---------I---------I---------I
          2.0000       I       42.826      I       85.80
   1, I100I        21.338              64.313

Raw Fixed Five
```

```
0.40000    I                              11 1 1 1  1
           I      1
           I
           I      1
           I     222                   111
0.31980    I     2 111
           I     1 1 1
           I         1         1111111
           I    2
           I          11       1
0.23960    I  1 1     11     1 1
           I
 3, ST1020 I    4       11   1
---------- I  6         1 11
           I 7 3
0.15940    I   3
           I 2            1
           I   1
           I
0.95238E-01I 4
             I---------I---------I---------I---------I
          2.7619       I       29.283      I       57.20
   1, I100I        15.325              43.242

Raw Moving Average Twenty
```

*Fig. B.3. Scattergrams: Ideal 100 val*

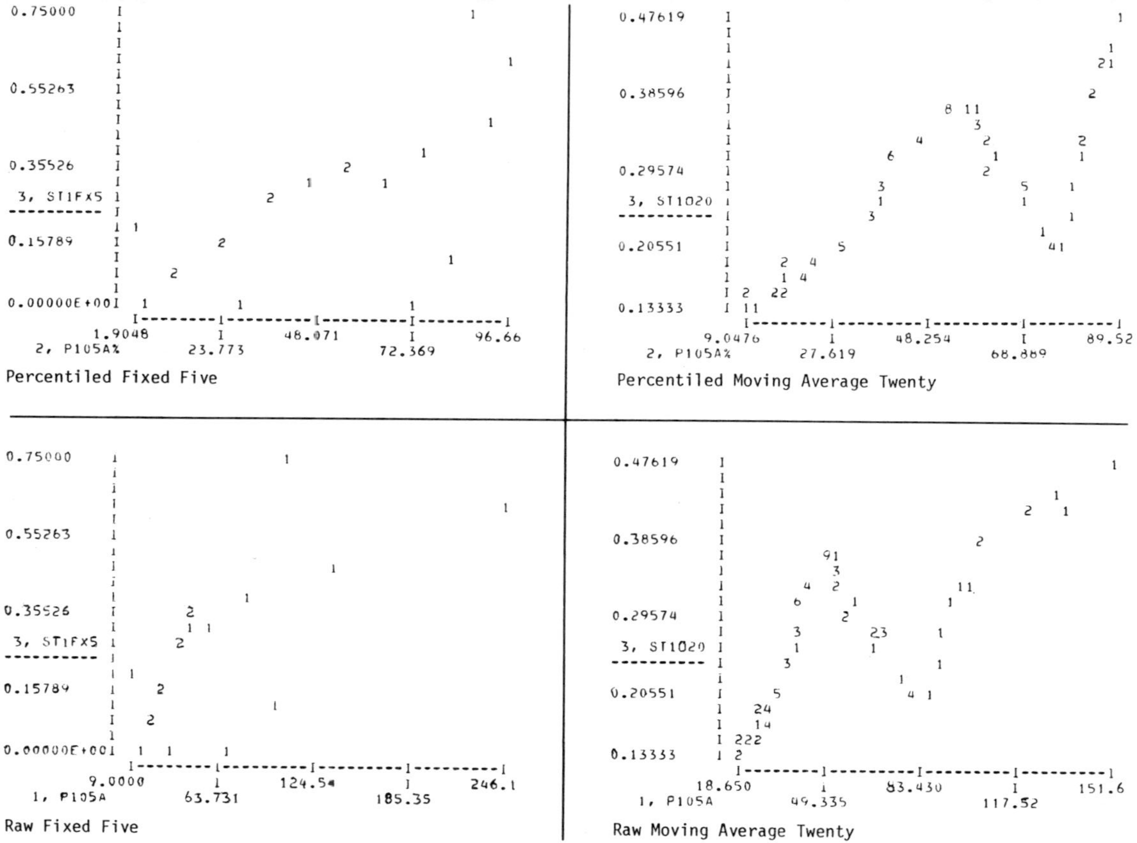

Percentiled Fixed Five

Y axis (3, STIFX5): 0.75000, 0.55263, 0.35526, 0.15789, 0.00000E+00

X axis (2, P105A%): 1.9048, 23.773, 48.071, 72.369, 96.66

Percentiled Moving Average Twenty

Y axis (3, ST1020): 0.47619, 0.38596, 0.29574, 0.20551, 0.13333

X axis (2, P105A%): 9.0476, 27.619, 48.254, 68.889, 89.52

Raw Fixed Five

Y axis (3, STIFX5): 0.75000, 0.55263, 0.35526, 0.15789, 0.00000E+00

X axis (1, P105A): 9.0000, 63.731, 124.54, 185.35, 246.1

Raw Moving Average Twenty

Y axis (3, ST1020): 0.47619, 0.38596, 0.29574, 0.20551, 0.13333

X axis (1, P105A): 18.650, 49.335, 83.430, 117.52, 151.6

***Fig. B.4.*** *Scattergrams: Pakistan 105A*

wishes to interpret the scale shown on the independent variable in Figures B.2–B.4. It is not the original independent variable shown in Table B.1 for VAL and PER. It is calculated by taking the mean of the actual values for the people in that interval as defined in the fixed five and moving average twenty calculations. For example, in the Ideal 100 fixed five calculations in Figure B.3, the number 90 shown at the end of the scale represents the mean of the last four values on the VAL variable shown in Table B.1. This "raw" variable is not as skewed in its distribution as the actual variable, but it contrasts enough with the percentiled variable to permit comparisons of their effects.

The scatter diagrams are complemented by least-squares regression done on the same data. Table B.3 shows a variety of regression results that I shall briefly describe and discuss.

Note first that the table provides results for both the raw and percentiled independent variable, so that the effect of the change in scale on regression results may be observed.

Six equations were estimated for each data set, and the display of summarizing statistics permits the reader to seek patterns in these results. Virtually all attempts to add the "x" term as the fourth term in the fourth-degree equation were unsuccessful due to the tolerance limits of the program being used.

Opinions vary on the degrees of freedom appropriate to the fixed five equations. The fixed five transformation replaces the 100 original data points in Ideal 100 with 20 averaged data points (and makes corresponding changes in Pakistan 105). Some would advocate calculating degrees of freedom based on N = 100, some calculation based on N = 20.* The unweighted results shown in the table use N = 20, the weighted ones N = 100. In the moving average twenty calculations N = 81. In all the results shown in the table $R^2$ is adjusted for degrees of freedom; that is, it can, and does in some cases in the table, go down with the addition of terms to the equation.

The table shows at least three things that are important in the present context. First, some forms of data transformation combined with polynomial regression analysis will yield a lower $R^2$ with a polynomial equation even when curvilinearity is clearly and explicitly built into the data (see especially the fixed five, unweighted, Ideal 100 results).

Second, it is possible to use polynomial regression to get substantially closer (higher $R^2$) fits than are possible with linear regression applied to both the simulated (Ideal 100) and the natural (Pakistan 105) data sets (see especially the moving average twenty results and the fourth-degree equations).

Third, many factors that apparently produce differing regression results are characteristics of polynomial regression analysis that are not encompassed by the theory. This is especially true of the differences

*Charles Lave, who provided invaluable assistance in the development of the regression analysis, considers it an open argument.

TABLE B.3. *Regression Analysis with Continuous Dependent Variable*

| DATA SET | Fixed Five, unweighted | | Fixed Five, weighted | | Moving Average Twenty | |
|---|---|---|---|---|---|---|
| Equation | $R^2$ | t Ratio[a] | $R^2$ | t Ratio[a] | $R^2$ | t Ratio[a] |
| IDEAL 100 PER (raw and percentiled are identical) | | | | | | |
| $x$ | .241 | 2.7 | .276 | 6.2 | .567 | 10.3 |
| $x+x^2$ | .204 | 1.1 ___ | .278 | 2.6 1.1 | .612 | 5.3 3.2 |
| $x+x^2+x^3$ | .179 | 1.1 ___ ___ | .294 | 2.7 2.0 1.8 | .724 | 7.3 6.1 5.7 |
| $x^2$ | .199 | 2.4 | .235 | 5.6 | .478 | 8.6 |
| $x^2+x^3$ | .172 | 1.1 ___ | .249 | 2.6 1.7 | .537 | 4.7 3.3 |
| $x^2+x^3+x^4$ | .189 | 1.5 1.3 1.2 | .303 | 3.6 3.1 2.9 | .822 | 12.8 11.8 11.2 |
| IDEAL 100 VAL raw | | | | | | |
| $x$ | .198 | 2.4 | .214 | 5.3 | .409 | 7.5 |
| $x+x^2$ | .222 | 2.0 1.3 | .257 | 4.2 2.6 | .408 | 3.0 1.0 |
| $x+x^2+x^3$ | .195 | 1.5 ___ ___ | .262 | 2.8 1.7 1.3 | .505 | 5.0 4.2 4.0 |
| $x^2$ | .095 | 1.7 | .132 | 4.0 | .349 | 6.6 |
| $x^2+x^3$ | .143 | 1.8 1.4 | .209 | 4.0 3.2 | .353 | 2.5 1.2 |
| $x^2+x^3+x^4$ | .107 | 1.0 ___ ___ | .212 | 2.1 1.5 1.2 | .361 | 2.1 1.6 1.4 |
| IDEAL 100 VAL percentiled | | | | | | |
| $x$ | .326 | 3.2 | .330 | 7.1 | .520 | 9.4 |
| $x+x^2$ | .298 | 1.4 ___ | .339 | 3.3 1.6 | .534 | 3.8 1.8 |
| $x+x^2+x^3$ | .290 | 1.5 1.0 ___ | .369 | 3.5 2.6 2.3 | .604 | 4.9 4.1 3.8 |
| $x^2$ | .258 | 2.8 | .274 | 6.2 | .456 | 8.2 |
| $x^2+x^3$ | .245 | 1.3 ___ | .296 | 3.1 2.0 | .485 | 3.6 2.3 |
| $x^2+x^3+x^4$ | .253 | 1.5 1.2 1.1 | .353 | 3.9 3.3 3.1 | .679 | 7.9 7.2 6.9 |
| PAKISTAN 105 raw | | | | | | |
| $x$ | .221 | 2.5 | .342 | 7.4 | .409 | 7.5 |
| $x+x^2$ | .199 | 1.6 ___ | .342 | 3.1 1.0 | .423 | 3.5 1.7 |
| $x+x^2+x^3$ | .164 | 1.2 ___ ___ | .339 | 1.8 ___ ___ | .639 | 8.0 7.2 6.9 |
| $x^2$ | .130 | 2.0 | .286 | 6.5 | .342 | 6.5 |
| $x^2+x^3$ | .144 | 1.5 1.2 | .325 | 3.7 2.6 | .350 | 2.6 1.4 |
| $x^2+x^3+x^4$ | .144 | ___ [b] | .325 | | .497 | 5.5 5.0 4.9 |
| PAKISTAN 105 percentiled | | | | | | |
| $x$ | .197 | 2.4 | .303 | 6.8 | .492 | 8.9 |
| $x+x^2$ | .147 | ___ ___ | .303 | ___ 1.0 | .557 | 5.4 3.6 |
| $x+x^2+x^3$ | .230 | 1.8 1.7 1.7 | .359 | 3.2 2.9 3.1 | .602 | 4.8 3.6 3.1 |
| $x^2$ | .178 | 2.3 | .307 | 6.9 | .397 | 7.3 |
| $x^2+x^3$ | .127 | ___ ___ | .302 | ___ ___ | .491 | 5.1 3.9 |
| $x^2+x^3+x^4$ | .177 | 1.5 1.4 1.7 | .440 | 5.1 5.0 5.1 | .754 | 10.8 9.8 9.2 |

[a] A ___ indicates the t ratio is less than 1.

[b] The $X^4$ term does not add substantially to the equation with $X^2$ and $X^3$ alone, and is below the tolerance of the regression program used.

between the third degree ($x + x^2 + x^3$) and the fourth degree ($x^2 + x^3 + x^4$) equations. By, in effect, requiring specification of parameters that the theory is not intended to specify, the use of regression analysis can obscure answers to questions about which the theory does have something to say.*

These and other issues are so complex, especially given my limited understanding of regression analysis and the paucity of data that would permit further exploration of the characteristics of polynomial regression within the scope of this study, that I have not found a way to use regression analysis to advance the substantive goals of the study. Thus, I will present more straightforward descriptions of the continuous forms of the data below.

### *A Bivariate Descriptive Display of Data with Percentiled Independent and Continuous Dependent Variables for Stage 1*

Simple scattergrams can be used to show the relationships of the transformed variables. This is done below. Figure B.5 displays the percentiled independent variable and the moving average twenty dependent variable for every study in which the data permits continuous analysis.† Following the arguments presented in Chapter 6, the underlying independent variable is land area in every case except Kenya 540, U.S. 159, and U.S. 167 (wealth index is used for Kenya and net farm income for the U.S.). This form of display is exactly parallel to the percentiled, moving average twenty form in Figures B.2–B.4.

*Boyd 1979 uses data from the present study and regression analysis with orthogonal polynomials in an attempt to separate the relative contributions of different components of the theory.

†Of the original sixteen cases, five (Mexico 108, Mexico 193, Taiwan 159, Taiwan 237, and U.S. 316) are too lumpy in the original to permit continuous analysis. The nonproportional, stratified, random sampling used in Pakistan 350 made it too complex for straightforward transformation into the form used here.

INDIA 185A

0.52000 / 0.39412 / 0.26825 / 3, ST1020 / 0.14237 / 0.41667E-01

9.0551 / 27.680 / 48.375 / 69.069 / 89.76

2, I185A%

INDIA 412A

0.35211 / 0.26957 / 0.18703 / 3, ST1020 / 0.10449 / 0.38462E-01

12.015 / 30.302 / 50.622 / 70.942 / 91.26

2, I412A%

KENYA 540I

0.33333 / 0.29650 / 0.25967 / 3, ST1020 / 0.22284 / 0.19337

17.241 / 34.019 / 52.660 / 71.301 / 89.94

2, K540I%

PAKISTAN 221A

0.58537 / 0.48980 / 0.39424 / 3, ST1020 / 0.29867 / 0.22222

9.9548 / 28.542 / 49.194 / 69.846 / 90.49

2, P221A%

**Fig. B.5.** *Scattergrams; India 185A, India 412A, Kenya 540I, Pakistan 221A, Philippines 153A, Philippines 177A, U.S. 159I, U.S. 167I, India 676A, India 1,145A*

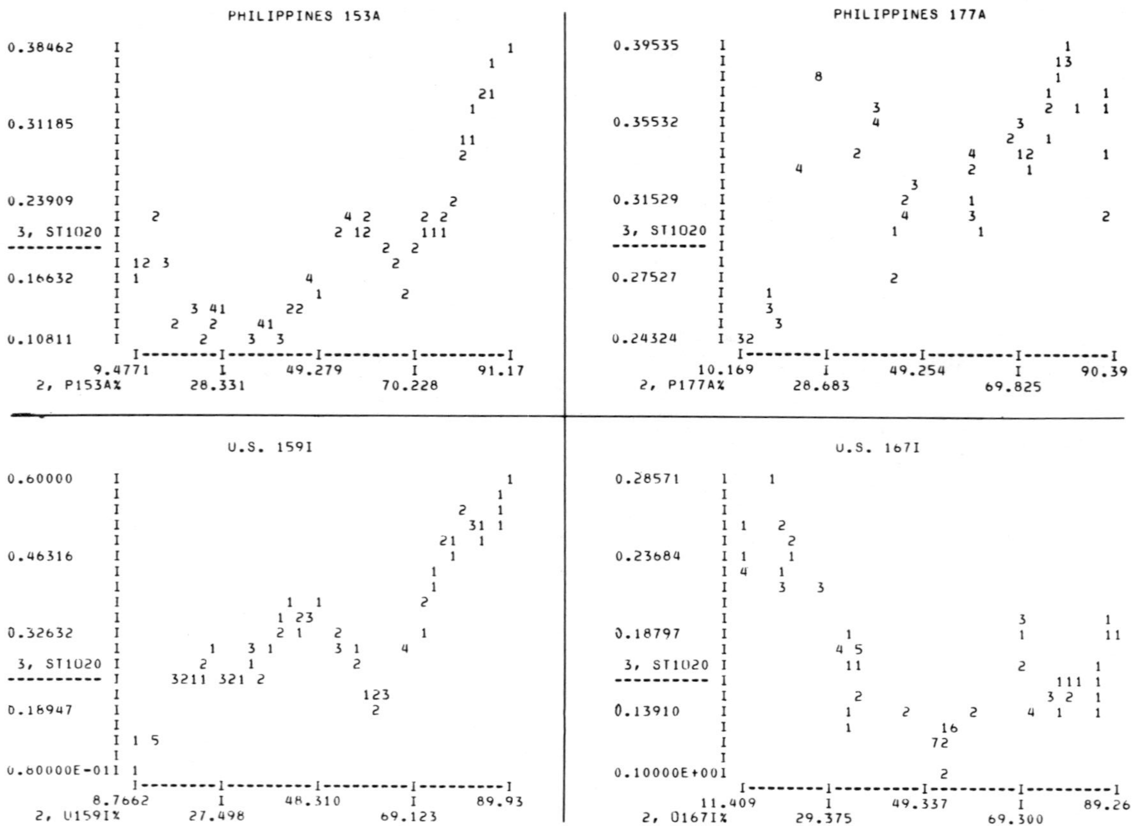


*Fig. B.5. Continued*

```
                          INDIA 676A
0.41135     I
            I                                              11
            I                                             21
            I
            I                                               2
0.33257     I                                             2
            I
            I
            I                                           2
            I
0.25380     I                                          2
            I                                         11
 3, ST1020  I                                      1  1
----------  I                           43           12
            I           4 411  32         123  14  3.
0.17502     I               2    1  5  1       1
            I      43       1
            I
            I
0.11200     I  3  4
             I---------I----------I----------I----------I
        9.1716         I      48.305         I        89.49
   2, I676A%        27.708                 68.901
```

```
                          INDIA 1145A
0.52444       I                                             1
              I                                             1
              I                                             1
              I                                            1
              I                                            1
0.40041       I                                           2
              I
              I                                        1222
              I                                        2
              I                                   1    2
0.27637       I                                   2223
              I                               23
 3, ST1020    I                             23
----------    I
              I                          212
0.15233       I                       5  14
              I                   31  3
              I              13  4  3
              I         1  23
0.53097E-01I  31
               I---------I----------I----------I----------I
         9.8339          I       48.949        I        90.12
   2, I114A%          28.362                 69.536
```

*Fig. B.5. Continued*

## *Appendix C*

# *The Analysis of Subpopulations*

Eight of the twenty-three studies were eliminated from consideration in Chapter 6 because their samples came from large universes that were thought unlikely to be communities of reference in the sense discussed in Chapter 3. Chapters 3 and 6 include statements of conditions under which population size is not an adequate guide to identification of a sociologically meaningful community of reference. This appendix will explore some cases that illustrate those statements. It will concentrate on four of the original studies, two of those that were eliminated and two that were not. Pakistan 221, discussed briefly in Chapter 6, will be used to illustrate the point that a sample taken from a large universe may, fortuitously, be representative of the smaller communities of reference within the universe. India 412 will be used to suggest that aggregation over many noncontiguous communities may obscure important differences. Mexico 108 and Mexico 193 will be combined to show that virtually adjacent communities can be importantly different.

Of the eight studies eliminated because they were originally sampled from a universe of more than 100,000 people, only Pakistan 221 and India 412 permit reanalysis in a manner that meets the criteria used for the fifteen cases retained. For each of these it is possible to define multiple subsamples of 100 or more farmers taken from universes of 100,000 or fewer people. It is not possible to approximate such subsamples for Pakistan 350, India 676,* India 1,145, or U.S. 316. Two cases, U.S. 341 and U.S. 423, are carryovers from Cancian 1967, and data were not available for reanalysis.

Of the fifteen cases retained, two (Mexico 108 and Mexico 193) come from nearly adjacent hamlets and can be recombined to form one sample from a universe of less than 10,000 people.

*In Cancian 1976b, three subpopulations including most of the farmers in India 676 are discussed, but the sampling universe for each of the three is more than 100,000 people.

The Pakistan 221 study, which confirmed both hypotheses using the 30/20 cut (see Table A.2), can be divided into two parts, Pakistan 105 and Pakistan 116. Each part is a *thana* (police jurisdiction) with a population of about 25,000, and each part also confirms Hypothesis A using the 30/20 cut (Tables C.1 and C.2). The thanas were selected in part for contrast within the district of 1.6 million people (see Table 3, also Rochin 1971); yet, as Tables C.1 and C.2 show, though Pakistan 105 is slightly poorer than Pakistan 116 in land area per farmer, they have roughly similar distributions on the independent variable. These tables were made up following the rules described in Appendix A. The distribution of HYV wheat adoption was also roughly similar in the two thanas; thus the cutting points for stages was the same. Figure C.1 shows the scattergrams for Pakistan 116 that parallel those for Pakistan 105 in Appendix B.

While the numbers are too small to support much more speculation, it seems clear that the Pakistan 221 sample, which must be seen as coming from a huge universe that cannot be assumed to be a community of reference, actually roughly replicates the distributions of the crucial variables in its subpopulations. In sum, a confirmation of the hypotheses by a sample taken from a large universe need not be a meaningless fluke. Without analysis of subpopulations, it is impossible to know what it means.

India 412 presents a different story. The sample of 412 includes two development blocks (N = 204 and 208), each with an estimated population of 100,000 people, and neither of these shows a strong pattern of confirmation of the hypotheses using crosstabular techniques of analysis. Table C.3 summarizes the results. Crosstabulations of the actual data are not provided here.

Each block sample is made up of censuses for four villages within the block (N = 31, 98, 30, 45, and 51, 33, 70, 54). Some of the villages are similar to others on the independent variable, some are substantially different. Although the one way analysis of variance shows the blocks to be different on the land area variable and not different on the income or adoption index variables, the breakdown to the eight villages shows differences among villages on all three variables. It happens that the cutting point for Stage 1 on the dependent variable is identical for each block and for the combined sample. It yields 21 percent in Stage 1 for India 204, 24 percent in Stage 1 for India 208, and 23 percent in the combined sample. This apparent homogeneity across the two development blocks is not supported by more detailed analysis. For the component villages the corresponding percent in Stage 1 using the cutting point that is common to the combined sample and each separate development block is: 36, 25, 3, 16, and 31, 52, 10, 19. Thus, the use of development blocks as universes may aggregate over very significant differences.

**TABLE C.1.** *Pakistan 105: Crosstabulations*

Area Q

| Count<br>Col % | .05 to<br>.30 | .35 to<br>.48 | 51 to<br>.76 | .81 to<br>4.05 | Row<br>Total |
|---|---|---|---|---|---|
| Remainder | 17<br>54.8 | 6<br>27.3 | 5<br>23.1 | 3<br>11.5 | 32<br>30.5 |
| Stage 2 | 9<br>29.0 | 10<br>45.5 | 12<br>46.2 | 11<br>42.3 | 42<br>40.0 |
| Stage 1 | 5<br>16.1 | 6<br>27.3 | 8<br>30.8 | 12<br>45.2 | 31<br>29.5 |
| Column Total | 31<br>29.5 | 22<br>21.0 | 26<br>24.8 | 26<br>24.8 | 105<br>100.0 |

Area T

| Count<br>Col % | .05 to<br>.35 | .40 to<br>.51 | .56 to<br>.91 | 1.01 to<br>4.05 | Row<br>Total |
|---|---|---|---|---|---|
| Remainder | 19<br>55.9 | 8<br>25.0 | 2<br>11.1 | 3<br>14.3 | 32<br>30.5 |
| Stage 2 | 10<br>29.4 | 13<br>40.6 | 11<br>61.1 | 4<br>38.1 | 42<br>40.0 |
| Stage 1 | 5<br>14.7 | 11<br>34.4 | 5<br>27.8 | 10<br>47.8 | 31<br>29.5 |
| Column Total | 34<br>32.4 | 32<br>30.5 | 18<br>17.1 | 21<br>20.0 | 105<br>100.0 |

Area D

| Count<br>Col % | .05 to<br>.25 | .28 to<br>.35 | .40 to<br>.40 | .46 to<br>.51 | .56 to<br>.76 | .81 to<br>1.05 | 1.11 to<br>4.05 | Row<br>Total |
|---|---|---|---|---|---|---|---|---|
| Remainder | 12<br>53.2 | 7<br>46.7 | 4<br>30.8 | 4<br>21.1 | 2<br>15.4 | 2<br>14.3 | 1<br>8.3 | 32<br>30.5 |
| Stage 2 | 5<br>26.3 | 5<br>33.3 | 5<br>38.5 | 8<br>42.1 | 8<br>6.15 | 8<br>57.1 | 3<br>25.0 | 42<br>40.0 |
| Stage 1 | 2<br>10.5 | 3<br>20.0 | 4<br>30.8 | 7<br>36.8 | 3<br>23.1 | 4<br>28.6 | 8<br>66.7 | 31<br>29.5 |
| Column Total | 19<br>18.1 | 15<br>14.3 | 13<br>12.4 | 19<br>18.1 | 18<br>12.4 | 14<br>13.3 | 12<br>11.4 | 105<br>100.0 |

TABLE C.2. *Pakistan 116: Crosstabulations*

Area Q

| Count<br>Col% | .05 to<br>.30 | .35 to<br>.51 | .56 to<br>.01 | 1.11 to<br>5.08 | Row<br>Total |
|---|---|---|---|---|---|
| Remainder | 9<br>28.1 | 9<br>31.0 | 14<br>50.0 | 7<br>25.9 | 39<br>33.6 |
| Stage 2 | 13<br>40.6 | 9<br>31.0 | 5<br>21.4 | 5<br>18.5 | 33<br>28.4 |
| Stage 1 | 10<br>31.3 | 11<br>37.9 | 8<br>38.8 | 15<br>55.6 | 44<br>37.9 |
| Column<br>Total | 32<br>27.6 | 29<br>25.0 | 23<br>24.1 | 27<br>23.3 | 116<br>100.0 |

Area T

| Count<br>Col % | .05 to<br>.35 | .38 to<br>.73 | .76 to<br>1.21 | 1.26 to<br>5.08 | Row<br>Total |
|---|---|---|---|---|---|
| Remainder | 11<br>29.7 | 9<br>33.3 | 14<br>45.5 | 7<br>29.2 | 39<br>36.6 |
| Stage 2 | 15<br>40.5 | 10<br>30.3 | 4<br>18.2 | 4<br>16.7 | 33<br>28.4 |
| Stage 1 | 11<br>29.7 | 12<br>36.4 | 8<br>33.4 | 13<br>54.2 | 44<br>37.9 |
| Column<br>Total | 37<br>31.9 | 33<br>28.4 | 22<br>19.0 | 24<br>20.7 | 116<br>100.0 |

Area D

| Count<br>Col % | .05 to<br>.20 | .25 to<br>.30 | .35 to<br>.40 | .46 to<br>.56 | .71 to<br>.91 | 1.01 to<br>1.26 | 1.32 to<br>2.02 | 3.04 to<br>5.86 | Row<br>Total |
|---|---|---|---|---|---|---|---|---|---|
| Remainder | 4<br>25.0 | 5<br>31.3 | 5<br>27.8 | 6<br>40-.0 | 6<br>40.0 | 8<br>44.4 | 5<br>35.7 | 0<br>0.0 | 39<br>33.6 |
| Stage 2 | 8<br>50.0 | 5<br>31.3 | 7<br>58.9 | 4<br>26.7 | 3<br>20.0 | 3<br>16.7 | 2<br>14.3 | 1<br>25.0 | 33<br>28.4 |
| Stage 1 | 4<br>25.0 | 6<br>37.5 | 6<br>33.3 | 5<br>33.3 | 8<br>40.0 | 7<br>38.4 | 7<br>50.0 | 3<br>75.0 | 44<br>37.9 |
| Column<br>Total | 18<br>13.8 | 18<br>13.8 | 18<br>15.3 | 15<br>12.9 | 15<br>12.9 | 18<br>15.5 | 14<br>12.1 | 4<br>3.4 | 116<br>100.0 |

```
 0.75000    I                                         1
            I
            I
            I                                1
            I
 0.55263    I                                       1
            I
            I    1                             1
            I
            I       2                1 1         1
 0.35526    I
            I         1      2   2       1
 3, ST1FX5  I 1
----------  I
            I           1                  2
 0.15789    I
            I
            I
            I
0.00000E+00 I   1                 1
             I---------I---------I---------I---------I
          2.5862       I       48.994      I       97.84
 2, P116A%          24.569             73.420
```

Percentiled Fixed Five

```
 0.54167    I                                       1 1
            I
            I                                      221
            I
            I
 0.47710    I                                        1
            I
            I
            I
            I
 0.41252    I                 2
            I    1         4               1     4
 3, ST1020  I                 1            1   2
----------  I             1  12     2         311
            I       6                     2
 0.34795    I     2                2
            I
            I 22 2      6      3          3
            I   2                2 14 4
 0.29630    I 1                         2
             I---------I---------I---------I---------I
          10.776       I       48.994      I       89.22
 2, P116A%          28.879             69.109
```

Percentiled Moving Average Twenty

```
 0.75000    I                                         1
            I
            I
            I            1
            I
 0.55263    I                   1
            I
            I  1          1
            I
            I  2     11     1
 0.35526    I
            I   122    1
 3, ST1FX5  I 1
----------  I
            I   1      2
 0.15789    I
            I
            I
            I
0.00000E+00 I 1    1
             I---------I---------I---------I---------I
          7.7143       I       187.62      I       377.0
 1, P116A           92.934             282.31
```

Raw Fixed Five

```
 0.54167    I                                1        1
            I
            I                            2  2     1
            I
            I
 0.47710    I                                   1
            I
            I
            I
            I
 0.41252    I       2
            I  1  4          1       4
 3, ST1020  I       1         1  2
----------  I     112  2        3 2
            I   6             2
 0.34795    I   2      2
            I
            I 42  6 3        3
            I  2      2 414
 0.29630    I 1              2
             I---------I---------I---------I---------I
          17.577       I       102.02      I       190.9
 1, P116A           57.578             146.47
```

Raw Moving Average Twenty

***Fig. C.1.*** *Scattergrams: Pakistan 116A*

TABLE C.3. *Hypothesis Tests on India 412 Subpopulations**

| | | Hypothesis A | Hypothesis B |
|---|---|---|---|
| INDIA 412 | 25 A | 0 | 0 |
| | 30 A | X | X |
| | 25 I | 0 | 0 |
| | 30 I | 0 | 0 |
| INDIA 204 | 25 A | 0 | 0 |
| | 30 A | X | X |
| | 25 I | 0 | 0 |
| | 30 I | 0 | 0 |
| INDIA 208 | 25 A | 0 | 0 |
| | 30 A | X | 0 |
| | 25 I | 0 | 0 |
| | 30 I | 0 | 0 |

*This table follows the conventions used in Table A.2.

The India 412 example suggests that combining noncontiguous populations may obscure substantial differences.*

The third example in this appendix illustrates how failure to separate virtually contiguous populations may obscure results that confirm the hypotheses proposed in this study. Mexico 108 and Mexico 193 (Apas and Nachig respectively) are nearly contiguous hamlets that together represent about 20 percent of the total population of the Maya Indian municipality of Zinacantan (Cancian 1965, 1972; Vogt 1969). The municipality population at the time of study was less than 10,000 people, and it was very distinct from adjoining municipalities. While the hamlets each have identities and important official and unofficial leadership roles, the municipality is overwhelmingly the dominant identity group from the outsider's point of view. It would seem reasonable to treat Apas and Nachig as appropriate parts of the same com-

*Each village was divided into ranks and stages in terms of its own distributions, and the resultant crosstabulations were combined into a single table in an attempt to avert the distortions introduced by aggregation. This combined table did not confirm Hypothesis A. Why it did not is another question. India 412 may represent a population of true negative cases. The village may be the inappropriate community of reference. India may be a negative case, unless the local caste system is taken into account (see Morrison et al. 1976).

TABLE C.4. *Mexico 301: Crosstabulations*

Area Q

| Count<br>Col % | .75 only | 1.5 only | 2.25 - 3.0 | 3.75 - 6.0 | Row Total |
|---|---|---|---|---|---|
| Remainder | 33<br>66.0 | 49<br>61.3 | 59<br>56.7 | 16<br>23.9 | 157<br>52.2 |
| Stage 2 | 8<br>16.0 | 19<br>23.8 | 26<br>25.0 | 31<br>46.3 | 84<br>27.9 |
| Stage 1 | 9<br>18.0 | 12<br>15.0 | 19<br>18.3 | 20<br>29.9 | 60<br>19.9 |
| Column Total | 50<br>16.6 | 80<br>26.6 | 104<br>34.6 | 67<br>22.3 | 301<br>100.0 |

Area T

| Count<br>Col % | .75 - 1.5 | 2.25 only | 3.0 - 3.75 | 4.5 - 6.0 | Row Total |
|---|---|---|---|---|---|
| Remainder | 82<br>63.1 | 34<br>56.7 | 32<br>51.6 | 9<br>18.4 | 157<br>52.2 |
| Stage 2 | 27<br>20.8 | 15<br>25.0 | 18<br>29.0 | 24<br>49.0 | 84<br>27.9 |
| Stage 1 | 21<br>16.2 | 11<br>18.3 | 12<br>19.4 | 16<br>32.7 | 60<br>19.9 |
| Column Total | 130<br>43.2 | 60<br>19.9 | 62<br>20.6 | 49<br>16.3 | 301<br>100.0 |

Area D*

| Count<br>Col % | .75 | 1.5 | 2.25 | 3.0 | 3.75 | 4.5 | 5.25 | 6.0 | Row Total |
|---|---|---|---|---|---|---|---|---|---|
| Remainder | 33<br>66.0 | 49<br>61.3 | 34<br>56.7 | 25<br>56.8 | 7<br>38.9 | 5<br>16.7 | 1<br>25.0 | 3<br>20.0 | 157<br>52.2 |
| Stage 2 | 8<br>16.0 | 19<br>23.8 | 15<br>25.0 | 11<br>25.0 | 7<br>38.9 | 17<br>56.7 | 2<br>50.0 | 5<br>33.3 | 84<br>27.9 |
| Stage 1 | 9<br>18.0 | 12<br>15.0 | 11<br>18.3 | 8<br>18.2 | 4<br>22.2 | 8<br>26.7 | 1<br>25.0 | 7<br>46.7 | 60<br>19.9 |
| Column Total | 50<br>16.6 | 80<br>26.6 | 60<br>19.9 | 44<br>14.6 | 18<br>6.0 | 30<br>10.0 | 4<br>1.3 | 15<br>5.0 | 301<br>100.0 |

*Fully detailed distribution. Not lumped by deciling rules. See last footnote in this appendix.

munity of reference, and, in fact, many important public activities take place at the municipality level (Cancian 1965). Nevertheless, the physical (geographical) differences between Apas and Nachig make it inappropriate to lump them for the sort of analysis in this study.

Apas is off the road (crops must be transported with pack animals) and close to the good agricultural land that Zinacantecos rent from non-Indians in the municipalities along the Grijalva River. Nachig is on the Pan-American Highway and further from the Grijalva River valley. These differences lead to substantial differences in response to the various government road-building and corn-buying programs that have provided opportunities for new agricultural practices (Cancian 1972). This "difference" is apparent when viewed from the point of view of new opportunities alone. That is, Apas and Nachig have apparently different rates of innovation. In fact, the relation of innovation to economic rank is the same in each hamlet (as analysis of Mexico 108 and Mexico 193 shows). But the strategies required by the different locations are substantially different (even though the hamlets are only a few miles apart and culturally homogeneous in contrast to people in surrounding municipalities).

Mexico 301 is formed by combining Apas and Nachig in one "sample," and using the rules stated in Appendix A to find the common cutting points.* As Table C.4 shows, Mexico 301 does not confirm Hypothesis A with any cutting point—despite the fact that its component parts, Mexico 108 and Mexico 193, both confirm Hypothesis A. The labels on these cases in Table A.1 show that mean farm size in Nachig is substantially greater than in Apas. Moreover, the cutting points for the stages are such that the combined cutting point includes no Apas farmers as innovators. The details of the local situation that make sense of these facts are included in Cancian 1972. For present purposes, I want only to point out how crucial identification of the appropriate community of reference is. In this particular case it is not clear that the hamlet is a stronger sociological community of reference than the municipality. It is clear, however, that the hamlet is the appropriate community of reference for measuring farm size as an indication of rank, and for interpreting the dependent variable.

Each of these illustrations points to the importance of specifying community of reference in terms appropriate to local conditions.

*Except that the rules for decile cuts are not applied. Thus the detailed distribution on the independent variable is shown in Table C.4.

## *Appendix D*

# *Background Publications for Each Study*

The following list includes the publications named in Table 2, along with others relevant to interpretation of the data sets used in this study. It is meant to get the reader started in the background literature, and is by no means an exhaustive list of relevant publications.

| | |
|---|---|
| 1) India 185 | International Rice Research Institute 1975; Parthasarathy 1975 |
| 2) India 412 | Roy, Waisanen, and Rogers 1969 |
| 3) Kenya 540 | Almy 1974 |
| 4) Mexico 108 | Cancian 1972 |
| 5) Mexico 193 | Cancian 1972 |
| 6) Pakistan 221 | Rochin 1971 |
| 7) Pakistan 350 | Lowdermilk 1972 |
| 8) Philippines 153 | Pal and Polson 1973; Polson 1966; Polson and Pal 1955, 1957 |
| 9) Philippines 177 | Same as Philippines 153 |
| 10) Taiwan 159 | Lionberger and Chang 1968 |
| 11) Taiwan 237 | Lionberger and Chang 1968 |
| 12) U.S. 159 | Lionberger 1963; Lionberger and Campbell 1963; Lionberger and Chang 1965; Lionberger and Corpus 1972 |
| 13) U.S. 167 | Same as U.S. 159 |
| 14) India 676 | Kivlin, Fliegel, Roy, and Sen 1971; Rogers, Ascroft, and Roling 1970; Roy, Fliegel, Kivlin, and Sen 1968 |
| 15) India 1,145 | Gartrell 1974, 1977 |
| 16) U.S. 316 | Bharadwaj and Wilkening 1973, 1974; Gartrell, Wilkening, and Presser 1973; Leuthold 1967; Wilkening and Bharadwaj 1966, 1968; Wilkening and Guerrero 1969 |
| 17) Japan 91 | Lindstrom 1958, 1960 |
| 18) Mexico 93 | Cancian 1967, 1972 |
| 19) U.S. 173 | Fliegel 1957 |
| 20) U.S. 251 | Gross 1942, 1949; Ryan and Gross 1943, 1950 |
| 21) U.S. 252 | Marsh and Coleman 1954, 1955 |
| 22) U.S. 341 | Wilkening 1952 |
| 23) U.S. 423 | Dean, Aurbach, and Marsh 1958 |

# *References*

Almy, Susan W.
1974 Rural Development in Meru, Kenya: Economic and Social Factors in Accelerating Change. Ph.D. dissertation, Stanford University. University Microfilms No. 74–26979. Ann Arbor: University Microfilms.

Bharadwaj, Lakshmi K., and Eugene A. Wilkening
1973 Canonical Analysis of Farm Satisfaction Data. *Rural Sociology* 38 (2): 159–173.
1974 Occupational Satisfaction of Farm Husbands and Wives. *Human Relations* 27 (8): 739–753.

Bickel, P. J., E. A. Hammel, and J. W. O'Connell
1975 Sex Bias in Graduate Admissions: Data from Berkeley. *Science* 187: 398–404.

Blalock, Hubert M., Jr.
1972 Social Statistics (2d ed.). New York: McGraw-Hill.

Boyd, John P.
1979 Three Orthogonal Models of the Adoption of Agricultural Innovation. Manuscript. University of California, Irvine.

Cancian, Frank
1965 Economics and Prestige in a Maya Community: The Religious Cargo System in Zinacantan. Stanford: Stanford University Press.
1967 Stratification and Risk-Taking: A Theory Tested on Agricultural Innovation. *American Sociological Review* 32: 912–927.
1972 Change and Uncertainty in a Peasant Economy: The Maya Corn Farmers of Zinacantan. Stanford: Stanford University Press.
1976a Social Stratification. *Annual Review of Anthropology* 5: 227–248.
1976b Reply to Morrison, Kumar, Rogers and Fliegel. *American Sociological Review* 41: 1089–1093.
1978 Practical Generalizations in Development Anthropology. Paper presented at the Tenth International Congress of Anthropological and Ethnological Sciences, New Delhi, December.
1979a Consuming Relationships (a review of Fred Hirsch's Social Limits to Growth). *Reviews in Anthropology*, Summer (forthcoming).

1979b A Useful Distinction Between Risk and Uncertainty. *Proceedings* of the Workshop on Socio-Economic Constraints to Development of Semi-Arid Agriculture, ICRISAT, February.

n.d. Risk and Uncertainty in Agricultural Decisions. *In* Agricultural Decision Making: Anthropological Contributions to Rural Development, Peggy F. Barlett, ed. New York: Academic Press (forthcoming).

Dittes, James E., and Harold H. Kelly

1956 Effects of Different Conditions of Acceptance on Conformity to Group Norms. *Journal of Abnormal and Social Psychology* 53: 100–107.

Dean, Alfred, Herbert A. Aurbach, and C. Paul Marsh

1958 Some Factors Related to Rationality in Decision-Making Among Farm Operators. *Rural Sociology* 23: 121–135.

Edwards, Ward, and Amos Tversky, eds.

1967 Decision-Making: Selected Readings. Baltimore: Penguin Books.

Fliegel, Frederick G.

1957 Farm Income and the Adoption of Farm Practices. *Rural Sociology* 22:159–162.

Gardner, Martin

1976 Mathematical Games. *Scientific American* 234: 119–122.

Gartrell, John W.

1974 Development and Social Stratification in India: The Structure of Inequality in Andhra Peasant Communities. Ph.D. dissertation, University of Wisconsin. University Microfilms No. 74–7469. Ann Arbor: University Microfilms.

1977 Status, Inequality and Innovation: The Green Revolution in Andhra Pradesh, India. *American Sociological Review* 42: 318–337.

Gartrell, John W., and E. A. Wilkening, and H. A. Presser

1973 Curvilinear and Linear Models Relating Status and Innovative Behavior: A Reassessment. *Rural Sociology* 38 (4): 391–411.

Goss, Kevin F.

1977 Consequences of Diffusion of Innovation. Paper read at the annual meeting of the Rural Sociological Society, September.

Gross, Neal C.

1942 The Diffusion of a Culture Trait in Two Iowa Townships. M.A. thesis, Iowa State University, Ames.

1949 The Differential Characteristics of Accepters and Nonaccepters of an Approved Agricultural Technological Practice. *Rural Sociology* 14 (2): 148–156.

Hirsch, Fred

1976 Social Limits to Growth. Cambridge: Harvard University Press.

Homans, George Casper

1961 Social Behavior: Its Elementary Forms. New York: Harcourt, Brace, and World.

International Rice Research Institute (IRRI)
1975 Changes in Rice Farming in Selected Areas of Asia. Los Banos, Philippines: International Rice Research Institute.

Kivlin, Joseph E., Frederick C. Fliegel, Prodipto Roy, and Lalit K. Sen
1971 Innovation in Rural India. Bowling Green, Ohio: Bowling Green State University Press.

Knight, Frank H.
1921 Risk, Uncertainty, and Profit. New York: Kelley.

Lave, C.
1973 Interactive Regression Program Using Free Field, Natural Language Control Commands. *Behavioral Science* 18 (March): 148.

Leuthold, Franklin O.
1967 Discontinuance of Improved Farm Innovations by Wisconsin Farm Operators. Ph.D. dissertation, University of Wisconsin. University Microfilms No. 67–12445. Ann Arbor: University Microfilms.

Lindstrom, David E.
1958 Diffusion of Agricultural and Home Economics Practices in a Japanese Rural Community. *Rural Sociology* 23: 171–183.
1960 Community Development in Seki-Mura: A Study Based on the Manual "Fact Finding with Rural People," with Emphasis Upon Diffusion of Farm and Home Practices in a Typical Japanese Rural Community. Urbana: Illinois Agricultural Experiment Station. Mimeo.

Lionberger, Herbert F.
1948 Low-Income Farmers in Missouri: Situation and Characteristics of 459 Farm Operators in Four Social Area B Counties. Columbia: *Missouri Agricultural Experiment Research Bulletin* 413.
1952 The Diffusion of Farm and Home Information as an Area of Sociological Research. *Rural Sociology* 17: 132–143.
1960 Adoption of New Ideas and Practices. Ames: Iowa State University Press.
1963 Legitimation of Decisions To Adopt Farm Practices and Purchase Farm Supplies in Two Missouri Farm Communities: Ozark and Prairie. Columbia: *Missouri Agricultural Experiment Station Research Bulletin* 826.

Lionberger, Herbert F., and Rex R. Campbell
1963 The Potential of Interpersonal Communicative Networks for Message Transfer from Outside Information Sources: A Study of Two Missouri Communities. Columbia: *Missouri Agricultural Experiment Station Research Bulletin* 842.

Lionberger, Herbert F., and H. C. Chang
1965 Comparative Characteristics of Special Functionaries in the Acceptance of Agricultural Innovations in Two Missouri Communities: Ozark and Prairie. Columbia: *Missouri Agricultural Experiment Station Research Bulletin* 885.

1968 Communication and Use of Scientific Farm Information by Farmers in Two Taiwan Agricultural Villages. Columbia: *Missouri Agricultural Experiment Station Research Bulletin* 940.

Lionberger, Herbert F., and Gary D. Copus

1972 Structuring Influence of Social Cliques on Farm-Information-Seeking Relationships with Agricultural Elites and Nonelites in Two Missouri Communities. *Rural Sociology* 37 (1): 73–85.

Lowdermilk, Max K.

1972 Diffusion of Dwarf Wheat Production Technology in Pakistan's Punjab. Ph.D. dissertation, Cornell University. University Microfilms No. 72–26348. Ann Arbor: University Microfilms.

Marsh, C. Paul, and A. Lee Coleman

1954 Communication and the Adoption of Recommended Farm Practices. Lexington: *Kentucky Agricultural Experiment Station Progress Report* 22.

1955 The Relation of Farmer Characteristics to the Adoption of Recommended Farm Practices. *Rural Sociology* 20: 289–296.

Merton, Robert K., and Alice S. Rossi

1957 Contributions to the Theory of Reference Group Behavior. *In* Social Theory and Social Structure (rev. and enl. ed.) by Robert K. Merton. Glencoe, Ill.: Free Press.

Miller, S. M., and Pamela Roby

1970 The Future of Inequality. New York and London: Basic Books.

Mischel, Walter

1977 On the Future of Personality Measurement. *American Psychologist* (April): 246–254.

Mitchell, J. C.

1956 The Yao Village. Manchester: Manchester University Press for Rhodes-Livingstone Institute.

Morrison, Denton E.

1973 Review of Change and Uncertainty in a Peasant Economy: The Maya Corn Farmers of Zinacantan, by Frank Cancian. *Contemporary Sociology* 2: 261–265.

Morrison, Denton E., Krishna Kumar, Everett M. Rogers, and Frederick C. Fliegel

1976 Stratification and Risk Taking: A Further Negative Replication of Cancian's Theory. *American Sociological Review* 41: 1083–1089.

Pal, Agaton P., and Robert A. Polson

1973 Rural People's Responses to Change: Dumaguete Trade Area, Philippines. Quezon City, Philippines: New Day Publishers.

Parthasarathy, G.

1975 West Godavari, Andhra Pradesh. *In* Changes in Rice Farming in Selected Areas of Asia. Los Banos, Philippines: International Rice Research Institute. Pages 43–70.

Pick, Franz
1968 1968 Pick's Currency Yearbook. New York: Pick.
1976 1975–76 Pick's Currency Yearbook. New York: Pick.

Pick, Franz, and René Sédillot
1971 All the Monies of the World. New York: Pick.

Polson, Robert A.
1966 The Impact of Change on the Villagers of the Philippines. *Indian Sociological Bulletin* 3 (3): 191–199.

Polson, Robert A., and Agaton P. Pal
1955 The Influence of Isolation on the Acceptance of Technological Changes in the Dumaguete City Trade Area, Philippines. *Silliman Journal* 2 (2): 149–159.
1957 Food Supply and Food Habits in the Dumaguete City Trade Area. *Silliman Journal* 4 (2): 107–113.

Rochin, Refugio I.
1971 A Micro-Economic Analysis of Small Holder Response to High-Yielding Varieties of Wheat in West Pakistan. Ph.D. dissertation, Michigan State University. University Microfilms No. 72–16502. Ann Arbor: University Microfilms.

Rogers, Everett M.
1962 Diffusion of Innovations. New York: Free Press.
1976 New Product Adoption and Diffusion. *Journal of Consumer Research* 2: 290–301.

Rogers, Everett M., with Joseph R. Ascroft and Niels G. Roling
1970 Diffusion of Innovations to Peasants in Brazil, Nigeria, and India. East Lansing: Department of Communication, Michigan State University. Mimeo.

Rogers, Everett M., with F. Floyd Shoemaker
1971 Communication of Innovations: A Cross-cultural Approach. New York: The Free Press.

Rogers, Everett M., with Lynne Svenning
1969 Modernization Among Peasants: The Impact of Communication. New York: Holt, Rinehart and Winston.

Rogers, Everett M., and Patricia C. Thomas
1975 Bibliography on the Diffusion of Innovations. Ann Arbor: Department of Population Planning, University of Michigan. Mimeo.

Roy, Prodipto, Frederick C. Fliegel, Joseph E. Kivlin, and Lalit K. Sen
1968 Agricultural Innovation Among Indian Farmers. Hyderabad-30, India: National Institute of Community Development. Also available as pp. 111–211 of Kivlin et al. 1971.

Roy, Prodipto, Frederick B. Waisanen, and Everett M. Rogers
1969 The Impact of Communication of Rural Development: An Investigation in Costa Rica and India. Paris: UNESCO; and Hyderabad-30, India: National Institute of Community Development.

Ryan, Bryce, and Neal C. Gross
1943 The Diffusion of Hybrid Seed Corn in Two Iowa Communities. *Rural Sociology* 8 (1): 15–24.
1950 Acceptance and Diffusion of Hybrid Corn Seed in Two Iowa Communities. Ames: *Iowa Agricultural Experiment Station Research Bulletin* 372.

Sailer, L. D.
1977 Programs for Intervalizing Data. Ancillae Scientiarum 4. Irvine: University of California, School of Social Sciences.

Schmitt, Raymond L.
1972 The Reference Other Orientation. Carbondale: Southern Illinois University Press.

Simon, Herbert A.
1957 Models of Man. New York: Wiley.

Stanfield, J. David, and Gordon C. Whiting
1972 Economic Strata and Opportunity Structure as Determinants of Innovativeness and Productivity in Rural Brazil. *Rural Sociology* 37 (3): 401–416.

Turner, V. W.
1957 Schism and Continuity in an African Society. Manchester: Manchester University Press for the Rhodes-Livingstone Institute.

Van Velsen, J.
1964 The Politics of Kinship. Manchester: Manchester University Press for the Rhodes-Livingstone Institute.
1967 The Extended Case Method and Situational Analysis. *In* The Craft of Social Anthropology, A. L. Epstein, ed. London: Tavistock. Pages 129–149.

Vogt, Evon Z.
1969 Zinacantan: A Maya Community in the Highlands of Chiapas. Cambridge: Harvard University Press.

Wharton, Clifton R., Jr.
1968 Risk, Uncertainty, and the Subsistence Farmer: Technological Innovation and Resistance to Change in the Context of Survival. New York: Agricultural Development Council. Mimeo.

Wilkening, Eugene A.
1952 Acceptance of Improved Farm Practices in Three Coastal Plain Counties. *North Carolina Agricultural Experiment Station Technical Bulletin* 98.

Wilkening, Eugene A., and Lakshmi K. Bharadwaj
1966 Aspirations, Work Roles and Decision-Making Patterns of Farm Husbands and Wives in Wisconsin. Madison: *Wisconsin Agricultural Experiment Station Research Bulletin* 266.
1968 Aspirations and Task Involvement as Related to Decision-Making Among Farm Husbands and Wives. *Rural Sociology* 33 (1): 30–45.

Wilkening, Eugene A., and Sylvia Guerrero
1969 Consensus in Aspirations for Farm Improvement and Adoption of Farm Practices. *Rural Sociology* 34 (2): 182–196.

# *Index*

# Index